THE RELATIVE VALUE OF THE
PROCESSES CAUSING EVOLUTION

THE RELATIVE VALUE

OF THE

PROCESSES CAUSING EVOLUTION

BY

AREND L. HAGEDOORN, Ph. D

AND

A. C. HAGEDOORN—VORSTHEUVEL LA BRAND

WITH 20 FIGURES

Springer-Science+Business Media, B.V 1921

ISBN 978-94-017-4564-2   ISBN 978-94-017-4728-8 (eBook)
DOI 10.1007/978-94-017-4728-8
Softcover reprint of the hardcover 1st edition 1921

CONTENTS

___

___

INTRODUCTION.

THE origin of species, the way in which evolution has taken place and is still progressing, seems, even now, to remain an open question. So much so, that three absolutely diverse theories of evolution still find their adherents. These three main theories are so different, that it would seem as if the real nature of the process of species-formation must for ever remain a subject for speculation only, and as if good facts are wholly wanting.

A minority of Biologists adhere to Lamarck's theory of the inheritance of adaptative changes induced by the environment; some incline to the view of Darwin and Weismann, that natural selection on small, individual variations gradually changes species, and still others believe with de Vries, that new species spring into being spontaneously, by mutation, saltation. Judging from this diversity of opinion, it would certainly seem as if no important headway had been made since Darwin.

This is very surprising, when we see how in the last fifteen years Genetics has become established as an actively developing branch of science, and counts by hundreds the Biologists who are engaged in genetical experimental investigations. All the data thus acquired, should have their significance for an insight into the process of evolution, the origin of species. This reluctance on the part of Biologists in general, and of Geneticists specially, to make an attempt to correlate the facts so far obtained, and to find out whether they shed any new light on the problem how new species may originate, is caused mainly, we hope, by the very diversity of the three above-named evolution-theories. Their great diversity necessarily leads to the idea that theories of evolution must always be essentially speculative.

We hope that this failure of Biologists in general to take an interest in the evolution-problem, is mainly caused by their unconscious recognition of the one-sidedness of the above-named theories, that each tries to explain all evolution by one omnipotent agency to the exclusion of all other causes. The spirit in which Darwin attacked the problem is hard to emulate. Few dare to presume they have the required broadness of vision to try what Darwin attempted, and it is not without a feeling of hesitation that one undertakes even to begin work of this kind.

But at times the feeling will not down, that a great number of eminent and able Biologists utterly fail to understand in how far their results have value in themselves, and in how far they are but building-stones. Many do not seem to see, that, without a timely attempt to utilize their results in construct-ing a coherent whole of a higher order, much detail is utterly wasted, being without significance in itself. It is very evident that many fail to see the forest because of all the trees.

There is no real spirit of subordination of the different genet-ical investigations to the main problems of evolution, in the work of any but Bateson and a few others. What we need from time to time, is to pause, and try to see what all the others are doing, where they are going, what work is left undone, and where work is unnecessarily duplicated.

This book is written with the purpose of showing that con-ditions are not so hopeless, that, especially in the last decade, great advances have been made in our knowledge concerning everything pertaining to evolution.

The existence of Genetics as a science, which is being devel-oped in a systematic way, so, that the majority of the students continue and amplify the work of others, dates only from the last fifteen years.

Genetical questions have always interested all sorts of Phil-osophers; they have also interested Botanists and Zoölogists. But it is remarkable to note how few of these Biologists have interested themselves, like Darwin, in genetical questions for the sake of their general aspect. The first author after Darwin who approached Genetics in this spirit was Bateson. Nearly all the Biologists who have interested themselves in Genetics have been Zoölogists or Botanists first, and Geneticians in the second place. Because, until recently, there existed in Genetics no points of contact with other branches of Biological science, no guiding principles, the work of the older Geneticians stands isolated in a remarkable way. Each of them tried to construct for himself a complete conception of the mechanism of variation of the inheritance of different qualities from parents to children, and of the causes for the diversity of animals and plants which inhabit the earth.

Most of these older Geneticians have been either Zoölogists or Botanists, and we will see how the very fact that one is Zoölogist or Botanist exclusively can hinder one in the making of his generalizations. This drawback comes to light directly we compare Darwin's generalizations with those of his followers, who, with the exception of Bateson, did not, as Darwin did, have an open eye for the most diverse facts of variation, and the influence of selection in animals as well as in plants. For instance, if we compare the ideas about evolution of the Zoölogist Weismann and the Botanist de Vries.

There is one point in common to all the theories of evolution, excepting Darwin's, and that is, that each theoretician has always over-emphasized one point, one single link in the chain of processes which goes to the making of species, and has brought out this point as "the" cause of evolution. Just as Lamarck has given all his attention to adaptation, and has led himself to believe adaptation to be the cause of numerous processes with which we now believe it to be only remotely concerned, so has Weismann over-emphasized the power of selection to the exclusion of everything else. And de Vries, who believed he had witnessed a striking instance of mutation, spontaneous origin of species, has come to believe mutation to be the sole important cause of evolution.

We are fully aware of the fact that the writings of such extremists as Lamarck, Weismann, and de Vries have been very useful in bringing adaptation, selection, and mutation to the fore, and so stimulating discussion of the importance and the relative rôle of these different processes in the making of species. But we are, nevertheless, convinced that these striking theories, which strain to make one process look important to the exclusion of everything else, have had their day. Genetics has been tamed. It is no longer the field where theorists of fanciful and proselytic tendencies, war, and preach, and ignore, each other's facts, but a regular inductive science, which strives to take into account and correlate all the facts adduced by Zoölogists, Botanists, and other specialists, a science in which experiments are constantly devised to get light on doubtful issues. Genetics has definitely passed from the stage of the book to the stage of the periodical.

Geneticians nowadays, cannot continue to make a clever point and by padding out the importance of some hitherto little regarded truth, make believe, that they have discovered the only cause of species-formation. One of the curious effects of over-emphasizing a single process and trying to make it pass for the whole of evolution, is obviously, that it is always necessary to make far-reaching generalizations from a slender body of facts. Or it is even necessary to invent a purely hypothetical process, begging the question, to make the theory appear at all plausible.

Lamarck, wanting to convince the naturalists of his time, of the effectiveness of individual adaptation in changing species, had to assume the hereditary effect of such adaptations, and even to-day we see whole institutions vainly trying by an earnest application of their entire personnel, to adduce other than dialectic proof of such an effect, casually assumed as true by Lamarck, as an indispensable foundation of his theory of evolution.

Weismann, anxious to make us believe that selection on ordinary variation was the sole cause for specific diversity, as well as for specific purity, soon saw that this theory was incomplete so long as it did not explain the causes of variation. Differing from Darwin in his attitude toward facts and Science, he was so convinced of the power of selection, that he invented a purely hypothetical intra-cellular struggle between "determinants, germinal selection".

De Vries, wanting to believe, that what he had witnessed in Oenothera was the only way in which new species spring into existence full-fledged, was confronted with two main difficulties. On one hand there were numerous instances, in which, among cultivated plants and animals selection on small differences had a permanent and far-reaching effect; on the other hand, what he had witnessed in one small group of plants did not exist anywhere else. The first difficulty he ignored, and to meet the second one, he had to invent purely hypothetical "periods of mutability" and several minor hypotheses as to the internal causes of mutation.

Finally, those authors, such as Lotsy, who want to believe that crossing is the cause of species-formation, feel that crossing, even though it may be the only cause of heriditable variability, does not explain specific stability. There is only one course of action consistent with the wish to maintain crossing in the rôle of "the" cause of species-formation, and that is to deny variability within species. No Zoölogist would deny the existence of variation within species; the only way in which a Botanist can do it, is Lotsy's way, to take the term species away from what everybody else calls species, and to give it to those special species which exist in certain strictly autogamous plants, namely, the genetically pure groups of plants, which everyone else calls "pure lines" after Johannsen.

As long as it was possible, that finely wrought out hypotheses about inheritance and variation were flatly opposed in every important point, there did not exist a science of Genetics, even if there did exist Geneticians. The fact that two such absolutely opposed conceptions of the influence of selection on species formation can exist at the same time, as that of

Weismann and that of de Vries, clearly shows how insuffic-
iently Zoölogists and Botanists know each other's facts. It
further shows, that Genetics as a science ought not to be rank-
ed, and ought not to be taught in Universities as a branch of
Zoölogy or of Botany or of Agriculture, but should be a thing
by itself.

Darwin was chiefly concerned with evolution, and he tried
to make his theories about evolution fit all the facts, Zoölogical
and Botanical, of variation and heredity, which were known
in his day. The Geneticians after Darwin, have not continued
his work in the same spirit. Their theories have been chiefly
theories of heredity, and their ideas about evolution have too
often been generalizations of a small body of facts, either
Zoölogical or Botanical. No all-round Genetician, familiar
with the history of continual change of the different species of
domestic animals, would have generalized the facts observed
by de Vries in Oenothera into a theory of evolution as this
author did, neither would it be possible, that a Genetician
conversed with the results of selection in the lines of wheats
started by Louis de Vilmorin and the recent results of pure-
line-breeding, made a theory of evolution like that of Weismann.

Judging from the little interest of the latter-day Geneticians
in problems of evolution, it would seem as if the enormous
progress which Genetics as a science has made in the last fifteen years, did not help us to a clearer insight into just this
fundamental problem, how species have originated.

It seems worth while to us, to find out, in how far the new
facts which these fifteen years of genetical experiments have
given us, can help us further on the road taken by Darwin.
Darwin's views about evolution were in accordance with the
state of knowledge about variation and the effect of selection
of his period. It is time to see, whether it is not possible to
clear up some points which were dark to Darwin. It seems surprising, that after all these years of diligent work in the study
of Genetics, we must still start from Darwin's ideas about evo-

lution as a basis for contemporary enquiry into this subject, but the fact remains that after Darwin, no one has set forth a comprehensive theory of evolution worth the name.

We will try to show, how on some questions which were almost wholly dark to Darwin, new light has been shed by later facts. One of these questions, is that of the origin of variation. It is clear that no evolution, no production of new species is possible without variation of some kind. All the different theories of evolution start with variation. In the Lamarckian theory, variations are induced by the environment, and as the effect of this induction is thought to be directly transmittable, species are gradually evolved, one from the other, by a continual variation under the influence of the conditions under which the species live. In the theories of de Vries, two kinds of variation are distinguished, the small, individual variations of Lamarck, induced by the environment, and not, or rarely transmittable, and sudden variations of a more imposing kind, which have so appreciable connection with environmental conditions, and which are thought to be each the direct cause of the production of a wholly new and complete species.

In the Darwin—Weismann complex of theories, evolution is thought to be caused by a continued natural selection on small variations in all directions, the cause of which was a mystery to Darwin, and is sought by Weismann in an indirect action of the environment. Weismann's theory of evolution in its last phases of development was essentially like that of Lamarck.

Some of the weakest points in all these theories of evolution are these, that no sufficient account is made of variation, that different kinds of variation are not distinguished, and that the theories do not begin with the beginning, with the causes of variation.

Every theory of evolution must account for variation, it must give a plausible explanation of the causes of that variation which may be instrumental in species-formation, and in the second place it must account for specific stability. This second point is also present in all the important theories of evolution,

Lamarck thought that the stability of a species is obtained as soon as the reaction of a species to a changed environment is definitely accomplished, as soon as the species has come into a new state of equilibrium with its surroundings.

Darwin rather denied the stability of species, and thought that a moderate amount of variation is always present, and Weismann thinks of the final stability and purity of a species as the result of a long continued natural selection.

De Vries holds, that the unknown causes for the abrupt variation which produces a new species, imply a new stability. Species spring into existence suddenly, and they are stable from the very beginning of their existence.

Our task is obviously a double one. We will have to treat of variation, we will have to ask, how much the new facts which Genetics has so far given us, have taught us about variation; whether we can distinguish between essentially different kinds of variation, which of these kinds may be concerned with evolution and which kinds are not, and we will have to show what we now know about the causes of variation. On the other hand, we will have for task, to examine how far the new facts have taught us something concerning the way in which specific stability is attained. Both subjects merge one into the other; we will see how far our answers to questions (as to the nature of variation) help us to appreciation of the causes for specific stability.

HEREDITY

No book, purporting to give a review of the influence of experimental evidence since Darwin has had on our knowledge about evolution, would be complete without treating of the mechanism of heredity. Do we know more about the mechanism of heredity than Darwin did, and if so, how does our knowledge affect our understanding of evolution?

The view of Darwin, that heredity is a transmission from parent to off-spring of protoplasmic units, in some way determining the characters of the new individuals is still prevalent in a slightly modified form in very many theories of heredity. Weismann's hypothetical "determinants" are thought to be influenced by the characters of the parents and to "determine" the characters of the offspring; according to him there exist reciprocal relations between "germplasm" and soma.

De Vries' pangens, although called by the name Darwin gave to his hypothetical bearers of hereditary characters, are differently conceived. De Vries' idea of pangens is an "intracellular pangenesis" and he does not believe in a migration of pangens through the individual or in mysterious relations between the final qualities of an organism, its reactions upon the environment, and its germ-cells.

From all kinds of experiments on grafting, but especially from the results of Baur and Winkler's work on periclinal chimeras, we clearly perceive, that cells do not modify the inheritable constitution of neighbouring cells, and that therefore de Vries' conception of the mechanism of ·heredity is nearer the truth than Darwin's or Weismann's.

It matters little whether hypothetical determinants are thought to be diffused throughout the cell, or localized in the cell-nucleus, or even in a particular "locus" of a chromosome, all the hypotheses which see the inherited as "determinants" of characters or of organs, must have this in common, that they have to assume that these determinants are sometimes "latent" or inactive, namely in those instances, in which an individual which does not show a certain peculiarity that his parents have, nevertheless transmits it to some of its offspring.

With the growth of Biomechanics, the science of the factors of development and function, a quite different conception of heredity begins.

Wilhelm Roux distinguishes two fundamentally different sets of factors in the development of the organisms, determination-factors and realization-factors, the former constituting the "inherited" and the latter the environment. On different occasions we have tried to show that Genetics is essentially a branch of Biomechanics, concerning itself with a study of those factors in the development of an organism which are inherited, and we are still convinced that little progress can result from a conception of Genetics as a mere statistics of cross-breeding experiences.

The question whether the inherited is simple or multiple, has been long a point of discursive argument, and not the least benefit derived from Mendel's discovery has been, that his question has been definitely solved.

Those authors who take a Biomechanical view of inheritance instead of a morphological or a statistical one, believe, that numerous things are transmitted from parent to offspring, which each, by their presence in the cells, tend to influence one or more definite steps, processes in the development, whenever these steps are taken or those processes undergone. From a biomechanical standpoint it is clear that no special states, no latency or semi-latency or inactivity need be ascribed to those inherited things, which in a certain individual are not factors in the development and which nevertheless are transmitted by it to some of its children.

In every instance in which we investigate such "Latency",

such an inactivity, we see that the process to be influenced did not take place, that the developmental stage acted upon, was not passed by the individual. Throughout this book, we use the term Gene, as a neutral one, as proposed by Johannsen, for these inherited things. Genes can be factors in the development, inherited factors. All inherited factors are genes, but all the genes present in a germ, need not be factors in the development. A certain number of the genes will be transmitted regularly, while only occasionally, or not all, participating in the development of the individual. It is fundamentally wrong to use the term "Factor" for the genes, because it necessitates the assumption of an occasional latency. The presence, or absence of a certain gene, may determine a definite difference in the final qualities, but is inadmissable to speak of such a gene as of the "determiner" for that quality. All the other genes contributing to the developmental process which results in the character in question, could each and all in their turn be called its determiner. The use of the term "Unit-character" should, we think, be discontinued. The characters of an organism are not so many separate, separable things, they are all the result of the interaction of a great many factors, some inherited, genes, and some constituting the environment. There is no reason for the assumption, that occasionaly genes are present in a zygote in a state, which insures their inactivity. In so far as the coöperation or non-coöperation of a given gene to the development of an individual is determined in the zygote, is it determined by the combination of other genes present. It is clear, that this conception of "latency" of characters is fundamentally different from that of de Vries. We have every reason to believe now, that every gene is present in the zygote in the same state, that every zygote is a fresh beginning, that, in as far as an individual's character can be said to be determined in its germ, they are given in the combination of genes present not in peculiar states of some of them. An organism's qualities, characters, are the result of its development, as such they can not be said to be inherited. Its genes are inherited from its

parents only, not in some vague way from more remote ancestors. We believe the facts all tend to make the view more and
more untenable, that there is a distinction between germplasm
and soma. Everything points to it, that in essentials all the
cells of an organism, up to the moment of a formation of gametes are identical, that they gave at least an identical set of
genes, namely the set present in the zygote from which they
all descend. It is only a matter of technique which prevents us
to show that every body-cell, every somatic cell is a potential
producer of germ-cells. It is possible to show that under certain
circumstances one single epidermis-cell of Begonia or of Carda-
mine produces a complete plant, which in its turn is capable of
sexual reproduction, and the only possible explanation of the
formation of adventitious buds is, that in these instances somatic cells, which normally would not have germ-cells in their descendant, can be induced to produce branches capable of flowering and seeding. Whereas, it has not been shown to be
possible to trace the origin of germ-cells in animals to cells,
which normally would not have germ-cells in their descendants,
the work of Carrell and others on cultivation of somatic animal
tissue cells in vitro shows show these difficulties are mainly
a matter of technique.

An organism inherits, whatever it does inherit, exclusively
from its parents as part of its zygote. Such things as Häckel's
recapitulation theory, which states that the ontogenetic development is a recapitulation of the phylogenetic development of
the species, have now only historical interest. Phylogenetically
new characters, qualities which the members of a group possess
since a short number of generations, are not shown at some late
stage of development, but at the exact stage in development in
which the peculiarity in genotype on which the new character
is partly dependant, is exerting its influence. If, by cross-breeding, we produce a rumpless fowl or a waltzing rat, we do not
observe that in our new strain the chicks after, having developed normally, lose their tails, or that the rats begin to waltz
after they have developed normally at first, but the character

absence of tail will be shown by the embryo at a stage, on which the coöperation of the lacking gene would be necessary for a growing out of the vertebral column, or in the case of the rats, we can doubtless demonstrate the aberration at a stage of the development, at which normally the lacking genes would be necessary for a normal development of the internal ear.

The order in which an organism's characters are unfolded is one of structural necessity, it is given in the combination of the inherited factors present in the zygote and the non-inherited factors of the environment. If a church-tower is constructed out of materials donated by the parishoners, it matters not at all in what order these materials arrive. It may be that the gilded weather-vane and the tiles for the roof are amongst the first articles got together. First the stones must be used for the heavy foundations and buttresses. Only after the walls have reached a certain height can the lighter bricks be used for the upper part. Not before the walls are ready can the wooden super-structure be built, and the tiles will have to wait until the steeple is ready to receive them.

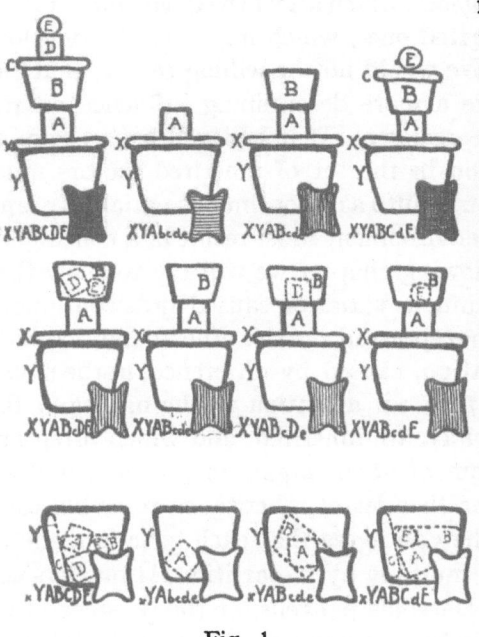

Fig. 1.
Diagram to illustrate the effect of presence or absence of some genes on the action of others, which modify the result of coöperation of the first ones to the development.

In some circles we meet with the remarkable question, whether the inherited or the environmental factors in the development are the more important. Especially does this discussion persist in the camps of the students of Eugenics. In mankind some would have us believe that the inherited only determined a child's character and peculiarities, others look upon men as born equal and different only according to the environment in which they grew up.

As soon as we leave the "determinant" view of inheritance, and place ourselves upon a biomechanical standpoint it becomes evident that the controversy is absurd. In so far as the development of an individual and therefore its characters can be said to result from separate factors in the development, these factors can be shown to be of two fundamentally different kinds, inherited ones, genes, which affect the development from within, and non-inherited ones, which influence the development from without. We would not be willing to say, that the inherited factors have a more determining influence on the final qualities of the organisms. It is believed by a group of authors, that variation in the set of inherited factors must necessarily translate itself into a discontinuous variability, and that variation in the environment must result in a continuous variation. In the following chapter we will try to show that there are cases of continuous variation caused by discontinuous differences in the genotype, and on the other hand, cases of discontinous modification, caused by differences in the environment. If now we turn our attention to the nature of the genes, those things which are inherited, and which often are factors in the development of the organisms, we see, that all the most diverse, older theories of inheritance resemble each other in one point, they all agree that each gene is made of protoplasm, and must multiply by bi-partition. It matters not whether we examine Darwin's pangens, or the pangens of de Vries, that are differently conceived, or Weismann's determinants or biophores, the biological molecules of Dewar and Finn, the bacteria of le Dantec and the brothers Simpson, or even the modern loci of Morgan and the cytologists, in every single case a gene is thought to be composed of protoplasm and to multiply by bi-partition.

When we take a biomechanical view of inheritance we have to look into this matter. To ascribe the qualities of the cells to qualities of genes, and then to turn round and state that these genes are protoplasm, is nothing but deferring the difficulty.

To ascribe a vital nature to the genes admits of explaining their variability; this must be the fundamental reason underlying the construction of these theories. As Dewar and Finn frankly state: These biological molecules have all the properties of living matter, including variability. And Weissmann's hypothesis of germinal selection, of a struggle for the available nourishment between the determinants, and de Vries' idea of latent and semi-latent and labile states of pangens could not be held, should the vital nature of the genes themselves be given up.

Since the time of Darwin, ever since Mendel's work got known, we have learned a great many facts about genes. To sum up: We know, that they are inherited from mother-cells to daughter-cells, but do not pass from cell to cell (Periklinal chimera's). We known that genes, which are inherited in only one gamete, will later be furnished to one half the number of gametes produced (Mendel's law), but that the influence upon the development in such a case is fully or approximately as great as in the case, where both gametes that make up the zygote contain the gene. We know, that the genes must have a nature which admits of their quantitative multiplication, but we also know, that the genes themselves are qualitatively stable and non-variable (Johannsen's law).

A vitalistic view of the nature of the genes certainly fits the facts, but whereas it is a theory that will work, it is not a theory that one can work with. The main new fact we know about genes is, what we have called Johannsen's law. It has been shown, conclusively we think, that inheritable variability is synonymous with genotypic impurity. For as far as a group of organisms contains some which have, and others which do not have certain genes, or some which are impure, (heterozygous) for one or more genes, this group has, what we want to call by the name of potential variability, and for so far as this goes, the group is amenable to change by selection or otherwise. But in those cases where we are sure, that the origin of the group insures a purity for one genotype, an absence of potential variability, selection has been shown to be ineffectual.

For this reason, no theory of the nature of the genes needs to make a provision for qualitative variability of the genes themselves. And this point was, we think, the only justification for a supposition that the genes are vital, protoplasmatic.

Protoplasm is clearly an emulsion, and it must be ultimately made up of a number of non-living substances, the combination of which makes it living. One of us has compared the attitude of the vitalist who reasons that every constituent of protoplasm which is an integral part of it, and which shows one or more properties of protoplasm, must itself be protoplasmatic and living, to the attitude of a philosophically inclined eater of plum-pudding, who would argue that the round, sweet things he could dissect out of his helping, and which looked like raisins could not be raisins, as he found them in his plum-pudding, and forming an integral part of it, they must consist of plum-pudding.

Quantitative propagation combined to qualitative stability is not exclusively a property of protoplasmatic bodies multiplying by bi-partition. Those chemical substances which have autokatalytical properties till both requirements, they propagate themselves, that is, suitable materials are ch .nged into a new substance under the influence of the presence of that substance. Also, they remain qualitatively unchanged. Some years ago one of us therefore published the hypothesis that genes are relatively simple chemical substances, non-living things, having autokatalytical properties.

Is this theory compatible with the facts known about genes and the action of genes? In the first place it does not admit of variation within the genes, or even within these groups of organisms which are known to be pure for all their genes, groups without potential variability. It is for this reason, that we have lately given so much attention to those cases in which it is claimed, that selection could modify the quality of genes.

Castle especialy professed a belief in the power of selection to change the nature of the genes. The controversy between Castle on one side, and Johannsen, East and ourselves on the other, does not concern a minor point, but it touches the very nature of the genes themselves. If it were true what Castle claimed, that selection can shift the influence one identical gene exerts over one identical developmental process, we would be no further in respect to an insight into the nature of the inherited than the authors of the vitalistic theories. It will be remembered, that one phase of the controversy concerned the effect of selection on the extent of pigmentation on Hooded rats. From the fact that all Hooded rats have one gene less than solid-coloured animals, Castle drew the conclusion, that Hoodedness was a unit-character, but he further concluded from the same fact, that all Hooded rats were geno-typically identical, and that therefore the proof that selection modifies the hoodedness, the extent of pigmentation, also proved that a gene had been modified in its quality.

To fully understand the case, it must be remembered, that hoodedness is recessive to solid colour, in other words that the Hooded rats are identical in one respect only, namely in the non-possession of one definite gene. It should be easy to understand, that the absence of such a gene does not imply purity for possible other genes which may influence the extent of pigmentation, but that the difference between animals with and without each of these genes need not be as marked as that between hooded and solid-coloured ones. It could very well have been impossible to demonstrate any of the possible genes, which make the difference between dark-hooded and light-hooded rats. As we did succed in demonstrating one of such genes, it may seem strange to allude to the possibility which confronted us at the start, of not being able to demonstrate any of the genes overlooked by Castle. But, it is obvious, that it is a thankless task, to repeat another experimentator's work over and

over, to look for facts, which the first man could have very
easily demonstrated himself.

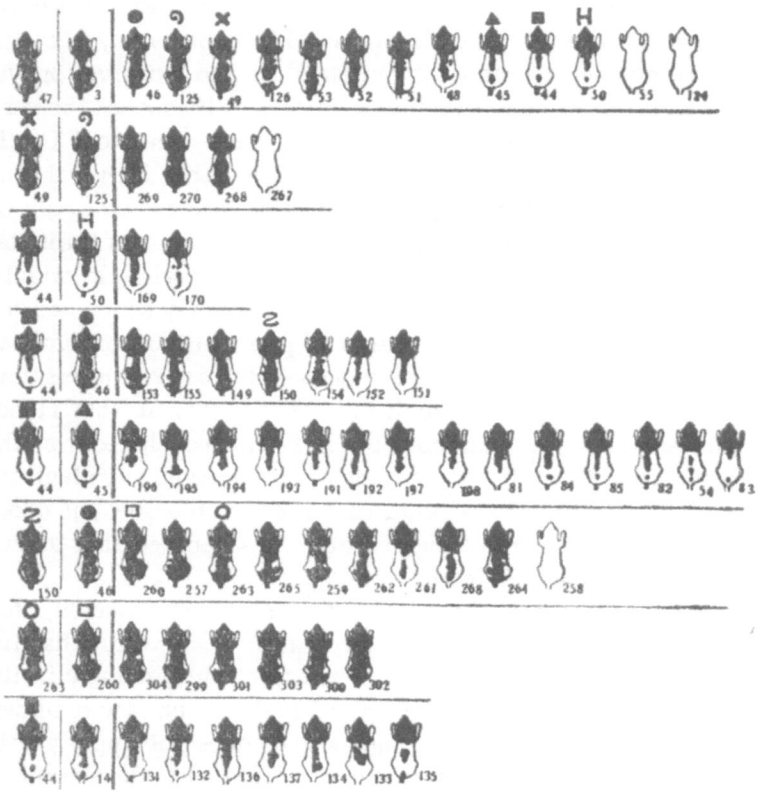

Fig. 2.

A few matings of hooded rats which show the existence of a gene,
which is present in dark-hooded and absent from light-hooded animals.
Light-hooded children from dark-hooded parents (e.g. 44 and 50, from 47
and 3) give no dark-hooded offspring. Dark-hooded may be homozygous
or heterozygous for the gene.

Young rats which are parents in any of the later matings shown, have
been given a distinguishing mark to facilitate finding their pedigree.

There happened to be in our hooded-rats animals with, and
others without one gene. The presence or absence of which,
exerted a marked influence on the extent of the pigmented

area in hooded-rats. We found, that two dark-hooded animals sometimes gave a minority of light-hooded young, and this was always the case when one of the parents of the animals had been light. Two light-hooded rats however, no matter what their ancestry, never produced any dark-hooded young.

A circumstance, which has certainly done much to strengthen Castle and his school in their belief in an effect of selection on the quality of the genes, is the over-distinctness of the difference resulting from a presence or absence of such genes, as were first studied by the Mendelians. The very fact that such genes, as produced a difference between black and white colour or between tall and dwarf stature were first studied, made it seem necessary to speak of such genes, as by their coöperation to the development produced a hardly appreciable difference in shade of black or a minute difference in stature, as of modifying factors. This unhappily chosen nomenclature, and the tendency to lump genes which happen to influence one and the same quality (no matter how, physiologically spoken) and speak of them as "polumeres," has created the impression which has certainly no foundation in fact, that such modification factors were the same old gene somewhat modified, and that sets of "polymeres" had originated by the splitting up of some one gene.

If the majority of the Geneticians had not approached Genetics from the side of Botany or Zoölogy, and had not started by observing a few striking inherited differences, but if they had happened to become interested in Genetics as in a study of those numerous factors in the development of the organism which are transmitted through the germ, we would not now find so many authors hampered by conceptions of unit-characters, and illusions about the purity of characters issuing from a cross.

Are the facts brought to light by the Cytologists compatible with my hypothesis, that the genes are relatively simple auto-katalytical substances? We know, that there are cases in which characters are inherited in an unusual way, that in some

crosses the inheritance of certain characters is exclusively
through the mother, and that no segregation occurs at the time
of gamete formation. The fact that in those instances, we are
concerned with the transmittancé of extra-nuclear material
(colour of the cotyledon of soy-beans) certainly points to the
cell-nucleus as to the organ responsible for Mendelian segregation. And, if inside the nucleus, we see such a very complicated
process as karyokinesis, and such significant modifications of
the process at the moment of production of germ-cells, all the
circumstantial evidence certainly points to the conclusion,
that the chromosomes are ultimately bound up with the process, which leads to a distribution of genes over cell-generations.
We hestitate to go further, and declare with so many authors,
to believe the chromosomes to be the bearers of heredity. In
these matters it is extremely difficult to see clearly, what is
cause and what is effect, what is of primary and what of secondary occurrence. Cytology, microscopic Technique and Mor-
phology are certainly ahead of Bio-chemistry of the cell, of
Micro-chemistry. But this admittance need not bring us to the
point, where we attach more importance to the more complete
morphological evidence than to the chemical facts, which are
obtained with so much more difficulty. The striking pictures
which Cytologists select for us from among their tens of thousands of stained sections of cells, certainly ought not to make
us sceptical concerning the chemical processes which we can
not see or make visible, but which assuredly accompany and
possibly cause the morphologic phenomena.

There is no incompatibility between the view, that the genes
are of a relatively simple chemical nature, and that they are in
some way localized on, or in, the chromosomes, but no one who
has read Trow's criticism of the evidence of "crossing over"
can prevent himself from being ever afterward rather sceptical,
when confronted with evidence for the localization in the Nth
locus on the Pth chromosome of a certain gene.

In this connection we may not forget, that the number of
genes we can study in any group of organisms is necessarily

very limited, that to study a gene, we must first happen to find individuals lacking it for comparison, and that for this reason only those genes which are not indispensable for an approximately normal life of the individual can ever be well studied.

We saw, that evidence is accumulating, showing that every somatic cell of a plant or animal has the genotype, the set of genes present in the original zygote. We have the evidence of regeneration in plants and animals and the fact, that in some plants it is possible to grow a complete plant like the mother-plant from one epidermis-cell. But, at the same time, we know that individual properties of different cells differ amazingly, morphologically as well as chemically. We would, to account for these facts, incline toward a belief that inside the nucleus a complete set of genes is somehow kept intact, whereas the cytoplasm of the different cells of one individual may be very different in different cells, to the point where one or two genes may be quantitatively preponderant. This view is a modification of de Vries' intracellular pangenesis with chemical substances substituted for vital pangenes. It is difficult to picture the way, in which a special kind of cell takes over a special function and prepares itself for that function so long as we conceive the genes as vital units, "determining" the cell's qualities. With the theory that the genes are autokatalitical substances, this differentiation becomes easier to understand.

. If we imagine, that a certain substance in the cells has some importance for the metabolism of a plant, and that this substance has autokatalytic properties, in other words, that it is a gene in this plant, we see how, wherever the constituents, the ingredients, for the formation of this gene enter the cells, they are assimilated and transformed into the combination, under the influence of this substance, the amount assimiliable proportionate to the amount of the gene present in the cytoplasm. It is evident, that after assimilating a quantity of the materials into this substance A, this cell or its daughter-cells become able to assimilate very much more of the same materials into A under the influe nce of a very much greater quantity of A. It is

very clear, that in this way cells and cell-complexes may specialize. From the very beginning of development, cells on the
outside of the embryo must in this respect be very differently
situated from cells, which are not in direct contact with the
surface. Cells at one pole of the embryo, may be under very
different conditions in respect to available food-supply. Even
differences of short duration may lead to lasting differentiation,
if the quantitative relation between genes in the different cells
induced by the difference, makes the cells react in a different
way to similar opportunities.

The phenomena of immunization find a ready explanation
on the theory, that the genes are autokatalisators. Just as we
can imagine how a cell can convert some materials with the aid
of a gene A into this substance A which may be a sugar, and
therefore, becomes able to convert still more of these same
materials into this sugar, so can we imagine how a toxic substance introduced in small quantities, can be used by some cells
in the upbuilding of one or more of the substances, which in
these cells are genes. A small quantity of a toxic substance, too
small to harm the life of the individual, is so converted into
the harmless substance of gene P, and after this transformation,
the individual is able to assimilate far greater quantities of the
toxic substance into P, with the aid of the quantity of P present,
before it has any harmful effect. It is possible, that what is
called the anti-toxin is an excess of a certain gene, which is
composed of "ingredients" taken mostly from the toxic substance introduced.

It is unsafe to let our fancy roam too far in these purely
speculative fields. But after this brief tentative explanation of
functional excitation and differentiation, we would like to
point out, how under the assumption that genes are relatively
simple chemical substances, protoplasm being an emulsion of
these substances, the difference of behaviour under selection
between uni-cellular and multi-cellular organisms, admits of an
explanation. In the first place, it is conceivable how a uni-cellular organism, a bacterium, gradually adapts itself to a new

substance in its culture-medium, through the fact that it can originally make use of very small quantities of it, transforming it or part of it into N under the influence of minute quantities of N present in its protoplasm as a gene.

We can imagine how, with the relative increase of N, and repeated cell-divisions, the capacity of the strain of bacteria for a splitting-up, or by an assimilation of the added substance, finally is greatly enhanced. It is possible in this way, to explain the process of adaptation of a clone of bacteria to a certain sugar, a change from a form which leaves the sugar intact or nearly so, to a strain capable of splitting it up. In the same way, we can vaguely picture how a bacterium which originally did not thrive as a parasite in an animal T. becomes in the course of many cell-generations adapted to live in this same species T.

A multi-cellular organism is less plastic. In the first place only a fraction of the number of its cells are in direct contact with assimilable substances; in the second place, it is probable that a complete set of genes is kept intact inside the nuclear membrane. I should judge that organisms without nucleus should be in a state of unstable genetic equilibrium. Autokatalytical substances, entering the organism from without, not only can increase in quantity within the protoplasm and contribute to the development in the same way as other genes, but in these organisms they are not at a disadvantage as compared with other genes.

But even if we compare more highly organized uni-cellulars with multi-cellular organisms, we see that there is a great difference in the course of heredity. Not only is the uni-cellular organism in intimate contact with its environment, but even those cells of the multi-cellular which do come into contact with the environment as a rule have no "future," that is to say, they have no germ-cells in their descendants. This leads us to the question of inheritance of effects of environment, and of spontaneous geno-variation, mutation.

When we get rid of the notion that genes are necessarily vital, complex things, which can vary, and can occur in latent or labile conditions, and substitute the hypothesis that genes are so many relatively simple chemical substances, we cannot conceive of spontaneous variation, variation that is not given by the potential variability, genotypic diversity of the group which presents it, as a function of the genes, as a process taking place from within, and we must have recourse to some explanation of spontaneous variation which sees the cause in the relation of the organism to its surroundings. De Vries assumes certain labile states of his vital pangenes, and ultimately he looks to the ambient circumstances to bring about these states, in which the pangenes occur. It is not difficult to see, how even if genes and the action of genes are the same in all organisms, spontaneous geno-variation may in its causation, be entirely different in uni-cellular from what it is in multi-cellular organisms. It is certainly a significant fact, that no authentic cases of a positive mutation in the higher plants or animals, are on record. Therefore, we may very well leave a discussion about the ways in which plants and animals can acquire new genes outside crossing, till the moment on which such an occurrence will be observed. On the other hand, in uni-cellular organisms, the taking up of a gene, new for the group, may not be an impossible rare process. It is for this reason, that we would like to warn against accepting instances of the effect of selection in apparently pure clones, of uni-cellular organisms as invalidating Johannsen's law.

If we accept the hypothesis that genes are relatively simple chemical things, and we search for an explanation of such cases of mutation as have been observed, that is to say of cases of the apparent loss of a gene, we must remember, that a distribution of a quantity of the gene in question over all the cells of the complete organism, is possible only, if the materials the "ingredients" of the substance, are available. To take an example. Let us suppose oxide-of iron, which is a substance with auto-katalitical properties, to be a gene, and a developmental factor in a family of plants. Only so long can the distribution of the substance be carried on, as iron and oxygen and oxide-of-iron are all three present. If the supply of iron fails, it can be imagined, that finally some cells may be produced lacking the oxide. If one of such cells happens to be a top cell, it may give rise to a bud, a branch, visibly different from the rest of the plant, a loss-mutation, bud-variation.

Until now, the experiments tried to induce mutation by external influences have been of a rather crude nature (Blaringhem, McDougal). We cannot see, that there was any working-hypothesis underlying such experiments which induced these authors to try particular injections, particular traumatisms. We think, that the best chance for success in experiments of this kind, lies in subjecting plants to conditions, which make it possible to regulate their available food-materials (water-culture) and to substract necessary ingredients from their food-medium for periods as long as is compatible with the health of the plants. We would not be surprised, if plants under such a treatment, produced buds, or branches, or seeds, lacking in one or more genes. There is no reason to assume a fundamental difference between bud-mutation and germinal mutations. It is evident that mutated branches give rise to mutated germ-cells, recessive mutations are perhaps more frequently produced spontaneously as bud-mutation than as germinal mutations.

Not only lack of an ingredient indispensable for a gene may be thought of as the direct cause of a mutation, but as different processes have different temperature-coëfficients, it is not difficult to imagine, how, in abnormally high or abnormally low temperatures, chemical processes, leading to the up-building the supply of a certain gene, may be temporarily suspended. We are here thinking of Tower's results with potatoe beetles.

Even if we think it highly improbable, that we will ever witness a progressive mutation, a spontaneous acquisition of a new gene by a higher plant or animal, we must remember, that the acquisition of a new gene by a new species as the result of cross-breeding need not be rare at all. Some authors seem to be willing to accept the occurrence of progressive mutations, wholly on the circumstantial evidence furnished by the existence of domestic species with new dominant characters. We must always remember, that such new dominant characters ·can be the result of a cross, even if the individuals crossed into the species do not have the character in question, for there is ample evidence for the fact, that very often genes are transmitted from parent to off-spring during long series of generation, without in any way being factors in the development, without contributing to the qualities of the individuals. The hybrid between two very similar, almost identical sub-species may have a very striking new dominant character.

The discussion as to the correct way to denote genes in Mendelian formulae, may be thought to be of very slight importance. Nevertheless, the discussion between Plate and one of us, emphasized the .difference in our conception of the relation between genes and characters. Mendel simply denoted the genes with wich he was concerned, by the first letters of the alphabet. Following the example of Baur we now do the same, preferring this to the system in use by very many authors, of calling a gene by a symbol, which calls to mind a character "determined" by it. In the first place, the mnemotechnic advantage is slight, for whereas it may be easy for Scandinavian and German readers to remember that S. denotes a gene, which black animals have more than yellow ones, readers of other nationalities would prefer to denote, the same gene by B, Z, N or I. But the chief objection to the system is, that it implies too much. If we observe that black animals have a certain gene more than yellows, there is no real blackness given in the gene. In different combinations of other genes, it will not produce a difference between yellow and black colour at all, but between cinnamon and agouti, or between cream and pearl-grey, or between champagne and lilac. And very often it will have no colour-modifying influence whatever. There is just as little reason to denote such a gene by a symbol recalling the colour black, as there would be to call sulphur B, because copper sulphate is blue.

We do not want to conclude this chapter without repeating

our protest against the too prevalent custom of denoting genes which have a similar effect by similar symbols. If three genes all tend to make the colour of a flower lighter, we do not think it better to denote them by L_1, L_2, L_3, than to use the symbols A, B and C for them. Physiologically, or biomechanically, there may be a very great difference in the way in which the presence of these genes tend to lighten the colour. Using the denotation L_1 L_2 L_3 heightens the impression that somehow these genes are components of one thing, that they have, at some former period, constitued one single gene.

————

VARIATION.

DARWIN in his theory of evolution started from the observation that all organisms are seen to vary in every character,
in every proportion, every function which is studied. Taking
for granted this variability, Darwin reasoned, that such small
differences as always exist between members of a single group,
must make these individuals better or less-fitted for the conditions under which they have to live.

And as it is manifestly impossible, for all the descendants of
all the individuals to find a place on earth, a portion only can
survice in each generation. Following this reasoning, and comparing the situation of species in nature with that of species of
domestic animals and plants, where selection brings about a
change of type and a reduction of variability, it seemed to
Darwin, that the survivors would always be those individuals,
which happened to be best adapted to their surroundings. If we
assume with Darwin, that those small differences, which we
can always observe as existing between the individuals of
every group, are hereditable, we can see how a continuance of
these two processes, on the one hand a variation, and on the
other hand a natural selection which tends to limit and direct
this variability, must result in a change. The group affected
must become more and more fit, and the direction of the gradual change is given by the conditions under which the group
lives.

Now this is all very clear and rather obvious, but two great
difficulties remain. Darwin's principle of natural selection
explains very nicely how a species, given a certain variability,
can develop some useful quality, and can become pure for this
quality, but it rather implies, that every quality for which a

given species in nature is found to be pure, must be materially useful. The theory does not show the way, in which a species may become pure for any quality which confers no advantage to the organisms which carry it. Even if we limit ourselves to domestic species, where selection has certainly been one of the main factors in the reduction of variability, we can explain how the white Wyandotte became pure for its white colour and yellow legs, and how the silver Wyandotte became pure for its peculiar feather-marking, but the explanation fails us if we enquire into the reasons which have made the white Wyandotte pure for a brown colour of its eggs, and why the silver Wyandotte always lays a pinkish egg with minute white dots, and another Wyandotte a white egg.

In the second place, the theory of natural selection, as proposed, by Darwin and elaborated by Weismann, does explain the way in which species may be thought to change if they vary, but it gives no explanation of the causes underlying this variation.

If we start from the assumption that a number of animals vary in height between four and ten, it is obvious enough that, if the tallest animals have a decided advantage, in a short time only animals of grade ten may survive. But in the first place, this does not explain why the animals did vary between four and ten; and further, it does not make it probable that the survival of animals of grade ten exclusively implies a variation of their descendants between, say, eight and twelve.

It can be seen how, given a certain small variation, this can be reduced by a suitable selection, and how ultimately the group can come to consist of extremes only. And this does not pre-suppose any knowledge about the causes of variation. But it remains a mystery why variation should continue, and how, as the result of the selection, it can now exceed the former limits.

No matter from which angle we look at the problem, we must see that the first question as to evolution has to be an inquiry into the causes of hereditable variation, and only if we have had this question answered can we expect a satisfactory answer to the second question: What causes, other than natural selection, can possibly reduce variability and render a group of organisms stable.

To Darwin, variability was a property of all living things, a natural phenomenon, which could possibly be enhanced by changes in the environment, by use or disuse of organs, but withal one, fundamentally simple phenomenon. This is one way of looking at variability, from the outside, and the term variability comes to mean nothing more than a statement of the fact that not all the individuals are identical.

From the moment we look closely at the differences which exist between the individuals of one group, we see that these differences cannot be all of the same kind, and different only in degree. If we observe a plot of wheat, we may observe that the plants which grow in a wet spot are taller than those on dry places of the same field. And, we perceive at once, that this difference is fundamentally of another nature, than the difference between an awned and an awnless plant which we find growing in that field. We can now perform a simple experiment, and sow four rows of wheat-kernels. One row from tall plants of the wetter spot, one row from the lower plants of the dryer portions, one row of seed from an awnless plant and a row of seed from awned plants. We will then probably observe that the first two rows will grow up alike, the seeds from taller plants will not give taller progeny than those from lower ones. But at the same time we will see, that the seed from awnless plants gives awnless plants and that the grain from awned plants reproduces awned off-spring. This shows, that the variability is different in the two instances. We see that in the last case, the difference between the awnless and the awned plants corresponds to a difference between the seeds these plants produce, but we see also that a difference between two plants does not necessarily imply a similar difference between the seed produced.

In studying variation and its causes, it is therefore necessary to ask, what is the relation between the quality of a seed and the qualities of the plants? What is the relation between the germ and the individual which grows from it? How does an organism come to be as we finally see it, how does it develop its characters?

If we contemplate an individual plant or animal, forgetting for the moment that it ever had ancestors, we understand that its present characters, the qualities which we see it have, are the result of its development, of the way in which it grew up from a germ. We see, that at every single moment of its development an embryo has its own qualities, which are continually changing in the course of this development. If we observe a morula develop into a blastula we witness a new quality which comes into existance, hollowness, or, we can say, we see the birth of a new organ, the cavity. This new quality of the embryo is the result of the development, of the multiplication of the cells of the morula and of the migration of the cells to the surface of the cell-agglomeration.

If now we remember that the organism *has* its ancestors, and that these ancestors, and the brothers and sisters, all develop into blastulas out of morulas, we see that there must be something common to the family, something common to all the individuals of the group which makes this sort of process happen in this way. We see that one or more of the causes of this development must be inherited, must be common to all the individuals of the group. But we also know quite a number of things which most certainly are also factors in the development of all these animals, and which also certainly are common to an enormous number of different sorts of organisms, factors which influence the development from without, such as the relation between salts in the sea-water in which the embryos develop, or oxygen, or gravity, which are of vital importance and whose coöperation to the development is absolutely necessary for the life of an individual. And there are others of relatively less importance, so that they may vary in wide limits, or even be lacking altogether, without seriously impairing the vitality of the individual.

Biomechanics was founded before Mendel's work was redis-covered, before it was known, that it would ever be possible to study the inherited developmental factors. For this reason, nearly all the work done by the founders of Biomechanics, has consisted in a study of the non-inherited, the environmental factors in the development of plants and animals. It has been abundantly illustrated that some of these factors may have a very definite action on certain stages of the development. Loeb found that a fertilized egg of a sea urchin, immediately after it had produced its fertilization-membrane, formed a second thin, close-fitting membrane, and that one of the things necessary for the production of this membrane was the presence of calcium-salts in the sea-water. If calcium was lacking this membrane was not produced. Normally, the two, or four, or eight cells of the young embryo are kept together by this membrane. If it is not produced, the two first blastomeres tend to drift apart, and each assume a spherical form. If the em-bryos are brought back into normal sea-water containing cal-cium, the membrane is formed, but as the two blastomeres are not flattened against each other, each of the two forms its own membrane, and further develops into a separate embryo. In this way, twins may be produced from normal eggs in nearly every instance. This case beautifully illustrates how a develop-mental factor, by influencing the development at a given in-stance, helps to determine the final qualities of the resulting organism.

As a typical instance of a developmental factor of another class, the ripening of corn-seeds may be described. There are, among other things, two kinds of corn which differ in that the seeds of one species are full of starch, whereas in the other in-stead of starch the seeds contain an abundance of sugar. If we plant the seeds of these two species side by side, and we com-pare the development of the seeds on two plants of different strains we see the following result. When the seeds are grown to their full size, the seeds of both plants are still soft and milky and they are full of sugar. If we cut off an ear of a plant of each kind and hang them up to dry in the sun, the water in the seeds evaporates, and as the envelop of the seed becomes too wide for the shrinking contents, it wrinkles, so that finally the seeds have a shrunken and somewhat glassy appearance like raisins.

If now we compare two ears of plants of different species, which we leave to ripen on the plants, we see that the ears do not behave in the same way. The seeds of the sugar-corn behave almost exactly in the way in which those in the cut ear did. The sugar concentrates, and as the water evaporates and the seed dries, it shrinks and becomes wrinkly and glassy. In the other plants, in the ripening seeds, the sugar is transformed into starch, and as a result of this process the seeds become hard, and on drying they remain so, and retain their shape. Therefore, we see that this difference between the hard starch-corn and the wrinkly sugar-corn, must be due to a developmental factor which at the time the seeds are ripening, is indispensable for a transformation of sugar into starch. It is not necessary to assume that this factor is in itself responsible for this transformation of sugar into starch, just as little as we need to assume, that calcium is alone responsible for the formation of the second membrane in the egg of the sea-urchin.

We must state it thus: for the formation of this second membrane in the sea-urchin, one of the indispensable factors is calcium, and for the transformation of sugar into starch in ripening corn-seeds, a certain heritable developmental factor is indispensable. We can never say that calcium causes the second membrane, that this membrane is determined by calcium. It may be possible to find another factor which is just as indispensable as calcium, for instance oxygen, and in the case of the corn we may not say that this one factor determines starch or starchiness. It is simply one thing which is necessary for the formation of starch.

In the example of the sea-urchin embryo, we are dealing with a developmetal factor of which the nature is known to us, and we know, furthermore, that it influences the development from without. In the corn-example we cannot lay our hands on the

factor we have been studying. It does not influence the plants
from without, for sugar-maize and starch-corn may be grown in
the same field in adjacent rows, and if pains are taken to exclude
crossing, the seeds of one row will only give sugar-corn-plants,
and the seeds in the other row will only produce starch-corn.
This proves, that here we are dealing with a developmental
factor which is transmitted through the germ, which is present
in the starch-corn, and there helped to produce the transformation of sugar into starch, and which is lacking in the germ of
sugar-corn. It must be a fundamentally different thing, an
inherited developmental factor. Its nature we cannot directly
infer, as we cannot isolate it from the plant which carries it.
One day it may be possible to find a chemical difference between the germ-cells of a sugar-corn, and of a starch-corn, but as
yet, no such difference has been detected, in spite of the efforts of Mr. Levallois. By comparing the two cases, we see that
the way in which calcium as a non-inherited developmental
factor affects the final result of the individual growing from a
germ, and the way in which in the case of our maize-example,
an inherited developmental factor affects the final qualities
of the individual growing from a germ, is comparable. Both
factors help to change the way in which the development pro-
ceeds by influencing a certain stage of it.

The study of the action of the different factors which influence the development of the organisms and so contribute to
their final qualities, is very much more difficult for the inherited factors than for the environmental factors. For, as it is
relatively easy experimentally to regulate the influence which
each environmental factor, temperature, salts, pressure, gravity, has on the developmental processes, we cannot do the same
thing for the inherited factors. As soon as a factor is of vital
importance for the life of the organisms to whose development
it contributes, we cannot study the way in which the development would proceed if the factor did not coöperate. And we can
therefore only study the action of such factors, if we can regulate the grade of their action, from optimum down to a mini-

mum which still allows of development. In this way we can study the effect of atmospheric pressure, and the action of light. But we know for certain that the inherited factors in the development of the organisms cannot be so regulated. They are either present in the germ, or they are absent from it. This has the result that only such transmittable developmental factors can be studied in their action as are not absolutely indispensable for at least a partial development. There is still another circumstance which hinders a study of this class of developmental factors. Such a factor can only be studied by a comparison of individuals, of which some have developed under the influence of it, and some without its coöperation. And as yet we have no means of eliminating these genetic factors. If in a species of plants or animals, all the individuals and all the germs produced contain a certain developmental factor, we can never study the action of this factor on the development, as the individuals which might have developed without it are not available for comparison.

We saw, that the different environmental factors which influence the development, can sometimes be different in intensity. In direct proportion to the variation of intensity in which a certain factor contributes to the development. The result, and therefore the final qualities of the organism may vary. Other developmental factors remaining constant, the amount of water available for growth may be directly related to a certain character of a population of plants, for instance to weight or height. The plants which have received more moisture may be taller than those which have had less water, and it is conceivable that in a field in which water is irregularly distributed, the plants are of varying height and all gradations of height may be present between that of the lowest and that of the tallest individual. Another example is heat. (Fig. 3). If we sow a number of seeds of the same pure strain, each in a little pot, and we put these pots in a row in a cold environment, then place a source of heat at one end of the row, we will see that the plants nearest this point are tallest, and those furthest away from it

will be lowest, and all the other plants will be of intermediate
height, varying from the tallest gradually down along the row
to the shortest plant. If we draw a line through the tips of

Fig. 3.

Above: a row of young plants of equal age, subject to greater or lesser influence of a scource of heat at the left. Continuous variation in
size.

Below: Similar row with glass-plate separating plants at the left from
those at the right. Discontinuous variation.

these plants, the resulting curve will be approximately smooth.
Other and different factors may influence the rate or the form
of development, hence the size of the young plants if they are
grown from seeds sown broadcast in a field. Some may receive
more moisture, others will have more shade, or more room than
the average. The result will always be, that the variation in
such a group is continuous, and that the variation of the resulting plant, in respect to such a character as height, when expressed graphically, will yield a typically normal Quetelet's
curve. It is variations such as these that are termed fluctuations. Ordinarily the effect of a variation in the non-inherited
developmental factors will be a continuous, fluctuating variability. But this is not always true. It may be that a non-inherited developmental factor varies continually, but that the
effect of this variation is not an equally continuous variation.
For instance, in the relation between such a factor and the

plant or animal which is influenced by it, there may be critical points. To take an instance, let us consider temperature. In the influence of temperature on the growth of plants, there is at least one critical point, the freezing-point of the sap.

Again, it may happen that a certain non-inherited developmental factor changes in intensity from a minimum to a maximum and back in long, regular periods, such as the average temperature in the course of a year. Some developmental process of an organism, and thus some characters of that organism may be gone through at two different, regularly re-

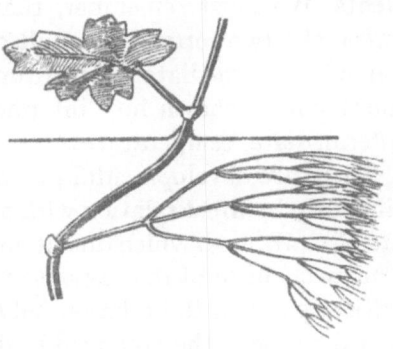

Fig. 4.

Discontinuous modification in the leaf-shape of the marsh-buttercup.

curring periods, which lie far apart, so that the extreme variants at each period do not reach the extremes of the other period. A few examples. The hair of certain animals is coloured only if the temperature and possibly other factors which change with the seasons, is above a certain minimum at the time of the moult. If such animals were moulting all the year round, they would gradually become darker-coloured in the spring and through the summer, and they would gradually lighten-up again in autumn and winter. As it is, there are animals which moult twice a year at definite seasons, once when the temperature is low, and once when it is high. Therefore a drawer of skins of such animals collected throughout the year exhibits a discontinuous variability. Very striking instances can be seen in the seasonal dimorphism of some species of bivoltine butterflies. Vanessa prorsa and Vanessa levana are strikingly different. A lot of Vanessa's comprising both spring and autumn-form exhibits discontinuous variability. In nature intermediates are not found. The springform passes the pupal stage in

winter, and the autumn-form in summer, and the difference
in result of the development may be explained by assuming
that some of the different developmental processes responsible
for pigmentation have mutually different temperature-coëfficients. We must remember, that the striking difference between the two forms is a result of the fact that no pupae develop in intermediate temperatures. The experiments of Dorfmeister have shown how intermediate types are produced in
intermediate temperatures.

There are developmental processes whose result varies continuously in direct relation with an important factor, but there
are also processes which do not admit of such a direct response.
The mechanism of the response may be such an one, that at a
critical point in the relation between the action of one factor
and the others, the organism begins to react differently. Some
phenomena, certain processes may by their very nature pro-
hibit continuous variation. For instance, meristic phenomena.
We do not know many definite facts about the causes for the
position of leaves on a stem, but we can see how the very nature of the obscure mechanism makes it impossible for a stem
to bear its leaves in any way intermediate between two definite ones. Thus it appears impossible for a plant which can
either bear its leaves in opposite pairs, or in whorls of three, to
have a stalk in an intermediate condition.

In Dahlia arborea, for instance, the better nourished stalks
bear their leaves in whorls of three, whereas the weaker stalks
and the smaller side-branches have their leaves in opposite
pairs. The cause of the difference, the thing which decides
that a branch will be of one or of the other kind, is here clearly
given by the non-inherited developmental factors. It is easy to
grow individuals of this Dahlia which are so ill-developed, that
not one stalk is produced with whorls of three leaves. And on
the other hand it is possible by appropiate methods which
cause the production of thick branches (pruning) to make the
lilac and some Fuchsias produce branches with the leaves in
whorls of three.

Numerous instances of discontinuous variation, resulting from a variation in the action of non-inherited factors of the development may be gathered. A beautiful example is the case of the twisted teazels reported by de Vries.

De Vries grew a strain of teazels, Dipsacus, which differed from the normal in their remarkable response to abundant nourishment. All the well-nourished, strong rozettes of this strain grow up into twisted individuals. All the weaker rozettes grow up normal. Intermediates are absent. That the difference is not dependent upon any difference in inheritable constitution, is shown by the fact that sowings of the seed of normal as well as of seed from twisted individuals always give the same mixture of normal and twisted plants, the relation between the two kinds being again decided by more or less nourishment. A great number of cases of "ever-sporting varieties" fall into this class, though obviously some of the cases of ever-sporting variability given by de Vries are fundamentally different, and are to be explained as rather complicated cases of segregation of inherited factors.

If we could study the influence of the different developmental factors of both types in an ideal way, so that all other factors could be kept constant, we would nearly always see, that the non-inherited developmental factors itself vary continuously, and that the resulting organisms would vary continuously in a corresponding way. And in the study of inherited factors, which as we know, cannot vary in intensity, but are either present or absent, we would see a discontinuous variation. We would always be able to distinguish those individuals, which had developed under the influence of our factor, from those to whose developement it had not coöperated. But of course in reality all the qualities of an organism are influenced each by so many different factors, which may vary independently, and in different ways, that the influence of the less important factors on the qualities of the organisms may be wholly obscured.

If we sow part of a homogeneous lot of seeds in a dry sterile field, and the remaining seed in a fertile well-watered place, we will, after the plants are harvested, still be able to differentiate the plants of the two lots with facility. But if we sow the seed in two adjacent fields of the same quality and in the course of the summer sprinkle a pail of water over one of the plots once, we shall never be able to sort out the plants of the two lots after they are once mixed at harvest.

With genetic, inherited factors, the case stands in the same way. If we have a lot of plants of which some have a certain inherited factor, necessary for pigmentation, whereas the others lack this, it will be easy at a glance to distinguish the plants of the two groups. If we mix the plants we are able to sort them again. The variation is discontinuous.

If we have a number of rabbits or mice, of which some are chocolate and some are black, because the latter have an inherited factor which the chocolates lack, we can still sort out the animals which are with or without this factor from a mixed lot. But the gap between the darkest, fullest-coloured chocolate, and the brownest, lightest-black is not nearly so large as that between a coloured plant or animal and an albino.

Another genetic factor in rodents, has still less influence on the development, and therefore, on the characters of the individuals. This is the factor which distinguishes fully-coloured from "fade" ones. (Fig. 5). If we compare two groups of animals, one of which has this factor and the other lacks it, we see an appreciable difference, but the variation is discontinuous only in certain families. In a group of black animals, the possession or lack of this factor makes so little difference, that the darkest, blackest animals without this factor may be appreciably blacker than the lightest and rustiest with it. In agouti animals however, the presence or absence of the same factor, appreciably changes the colour of the ventral side. A population of animals, or plants, which contains individuals with, and without, one genetic developmental factor, may, or may not show discontinuous variation. If the influence of the factor on the development is considerable, the gap between

the two sets of individuals may be so great that it is not filled
out by variations of both groups under the influence of the
environment in the widest
sense, and in other instances
it will not be large enough
to keep the variation from
being continuous. In a population of exclusively black
animals or exclusively hairy
plants, there may be individuals with and others without a factor which influences
colour or hairiness so little,
that the difference altogether escapes notice. There
are numerous genetic factors which we can never hope
to study, notwithstanding
the fact that they do in-
fluence the development.

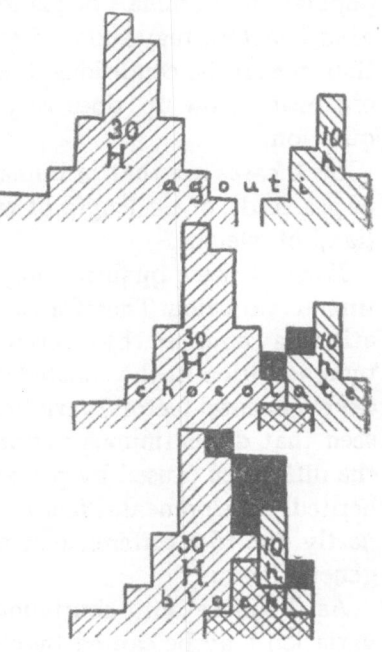

Fig. 5.

We have seen, that vari-
ation falls more or less clearly
into two different kinds,
continuous variation and
discontinuous variation. Continuous variation is obviously often the result of the fact
that the population which
shows it, is influenced by
several non-inherited en-
vironmental factors. But we
have seen that part of the
variability of such a group
may be due to the presence

Diagram to illustrate the shape of
variation curves in similar cases of
Mendelian 3: 1 segregation in the
off-spring of heterozygotes. The gene
for which the parents are heterozygous is the same in the three instances, but the remaining genotype is
different. Where the curves overlap,
individuals, represented by squares,
have been drawn in black. By pla-
cing the original variation-curves
closer together, the composite curve
becomes two-tipped, and a one-top-
ped variation-curve results long before the modes of the original curves
fall together.

or absence of inherited developmental factors. It may even
happen, that a number of genetic factors exist, each of

which tends to change the development a little in such a way
that the same quality is slightly changed by each of them. A
population of animals or plants which is impure for a number
of such factors, may show a typically normal continuous variation, even if the conditions, the environment, under which the
organisms grow up, does very little to change the quality in
question.

Such cases are that of human skin-colour, of the coherence
of pea-seeds, of ear-length of rabbits, of colour of wheat and
flax, of weight.

Each of these qualities may show a typically normal continuous variation. Therefore we can say, that continuous variation may be caused by a variation of the environment, but also
by a variation in the constellation of inherited, transmittable
developmental factors, variation of the genotype. And we have
seen that discontinuous variation is very often the result of
the difference caused by presence or absence of a gene, of inherited, developmental factors, but it may be often wholly or
partly due to a difference in the action of one or more non-
genetic factors.

As therefore both continuous variation and discontinuous
variation may be caused by a variation of the genotype, and
by a change in the complex of non-genetic factors, the environment, it is very clearly not possible to imagine that only
continuous variation or only discontinuous variation can be
concerned with evolution. Darwin thought that discontinuous
variation, the production of "sports" was a rare occurence,
and his theory of evolution is wholly concerned with the influence of selection on continuous variation. De Vries, on the
other hand, thinks, that continuous variation is wholly caused
by variations of the environment, and that only discontinuous
variation therefore can be the cause of evolution.

We have seen that there are essentially four kinds of variation:

A. Discontinuous geno-variation, discontinuous heritable
variation.

B. Continuous geno-variation, heritable variation.
C. Discontinuous non-heritable variation, modification.
D. Continuous non-heritable variation, modification.

We see that, if modification, the effect of the non-genetic factors, the influence of the environment, is not transmittable, we have in our search for the causes of inheritable variability to limit ourselves to geno-variation. But as we saw, it is not easy to distinguish geno-variation from modification. The idea that discontinuous variation and hereditable variation are synonymous and that continuous variation is modification, is surely erroneous. If we see a certain variation in any group of animals or plants, it is not at all easy to find out how much of the difference between the individuals is due to a difference of genotype, and how much to the effect of the conditions the developing individuals have encountered.

The question as to the causes of variation can be formulated as follows: What are the causes for the production of new combinations of genes? And to answer this question, it is necessary to review what we at the present moment know about genetic developmental factors. As yet we have always spoken of heritable developmental factors. We now want to use the ferm *gene,* for these factors as proposed by Johannsen.

It must here be stated clearly, that, whereas all heritable factors in the development of an organism are genes, we know about several genes which are in many instances not factors in the development of the individual which carry them.

We saw that some genes happen to influence the development of organisms in a striking way, for instance those genes which are links in a chain of factors which leads to pigment-formation. In individuals in which such a gene is lacking, there is no production of pigment, and therefore such individuals are strikingly different from others which have the same gene. But we have seen, that this difference between two individuals, with and without the same gene is not always so marked. We saw that some genes have so little influence in certain combinations, that it is hardly possible to distinguish individuals which have them from those which lack them. Our example of the gene which distinguishes "fade" from full black mice leads over to the case where a gene had no influence whatever. This same gene, which certainly has an appreciable influence on the colour of agouti mice, but very little on black ones, certainly has no influence on silver-fawn animals, as far as I could detect. All this shows, that a gene does not "determine" any character, but that it tends to influence certain definite developmental processes, and that, if these same processes do not take place for any reason, the gene has no influence. But this is something quite different from the assumption, that such a gene is transmitted in a state, which differs from the normal.

The gene itself is not latent, or dormant, or weakened in any way, for it can remain without influence throughout a number of generations and be as active as ever as soon as it finds the reaction which it influences. If we see that normal chickens give a minority of off-spring with drooping tails, we may assume that these normal chickens have a gene more than those with the drooping tail. We can imagine how this gene, which with other things seems to be necessary to make an otherwise drooping tail grow erect, can be transmitted even if the tail of all the chicks is cut off, or if the tail itself is wanting for some other reason. One of us observed a brood of young chicks of which the father was rumpless and the mother had a drooping tail, and which grew up with normal, erect tails.

The naïve idea, that every character or every organ has its own "determinant", makes it necessary to assume that in certain cases in which the characters or organs were not forthcoming, whereas the "determinant" was present (proved by crossing-experiments) this determinant was dormant or latent. I look upon the characters of an organism as upon the result of the whole development, and upon the development as the result of an enormous number of different factors of different kind.

For the production of bread in a bakery, a number of factors are necessary. We need a number of materials, flour, yeast, fat, water, salt, we need such things as the heat of the oven, and we need the skill of the baker. If any one of these important factors is lacking, bread cannot be produced. If there is no water, there will be no bread. If the baker is drunk, the result will be the same. It cannot be said that any of these necessary links in the chain of factors is "the determining factor." All the other factors being given, the last link completes the chain, and therefore in that particular instance can be said to determine the result.

The production of such a complex organ as a chicken's tail must be the result of a development by the interaction of quite a number of different factors, quite as diverse as the lot of factors in the production of bread. But some, on seeing that the difference between a normal and a rumpless fowl is sometimes due to the presence or absence of one single gene, will look upon this gene as upon "the determiner" for the tail. It is clear, that if we could succeed in getting hold of this particular gene, and putting it into the germplasm of some other animal without a tail, such as a hedgehog or a cavy, or a sea urchin, we could not very well hope to see it develop a chicken's tail.

If in our example of the bakery, eight different factors are absolutely indispensable for the production of bread, each of these eight things may in its turn be the only missing link, and as such "determine" bread or its absence. But no amount of this "determiner" introduced into any other combination of things excepting into this combination of the other seven, will determine bread. No matter how much water one pumps into any ordinary building, it will not produce one loaf of bread.

We cannot rightly say, that either the lacking factor or the result are "latent." If the flour runs out, none of the customers of the bakery will at breakfast-time take very kindly to the assertion that the bread is there all the same, only in a latent condition, because excepting flour all the ingredients are ready. And the statement that an albino mouse which, when mated to blacks, will produce agouti young, carries the latent factor for agouti colour, or carries latent agouti will not be accepted by those people who have learned to look upon genes as upon potential factors in the development rather than as upon determinants for characters.

We may not even say, that some genes are more important than others, or that, of a certain number of genes one is the determining one and the others are modifying factors. In many animals we have come to know a certain gene, which is necessary for any production of pigment. If this gene lacks, the animal growing from the germ will be an albino. The difference between a black and a grey animal on one hand, and that between a black and an albino is not of the same magnitude. But we may never think, that the gene which distinguishes coloured from albino is "stronger," than that which distinguishes black from agouti or chocolate from cinnamon.

The same gene which by its presence or absence "determines" the difference between a black and an albino mouse, will in other cases bring about the difference between a yellow and an albino, or between a pink-eyed lilac and an albino, a difference therefore, which is certainly not of the same magnitude as that between black and chocolate, or black and agouti.

We happen to possess pink-eyed white mice, which we took for albinos for some time after they came in our possession. They have, however, few minute spots of pale lilac-colour. A close search is necessary to distinguish between such mice and albinos, and we have one young mouse produced by this family whose status is undetermined. Only test-matings will show whether this animal has, or lacks, the same gene which produces the difference between our common glossy blacks and albinos.

Therefore, if we start breeding-work on colour of mice with some of these animals, we might declare that this gene was of minor importance, and that the result of its coöperation to the development, was not as great as that of the gene which distinguishes black from agouti.

The same gene may have a decidedly different effect on two divers combinations. In the brown rat, agouti animals are lighter-coloured than black, and this difference is due to the presence in agouti of one more gene.

The presence of this gene in yellow animals however makes the colour decidedly darker. In our house-rats the reverse is true. Here black is dominant over agouti and decidedly darker. The same gene, however, which causes the difference, makes the yellow animals, in which it is present, lighter.

If we conceive of the action of a gene, as of an influence on some process of development, part of a series of processes, we understand how the dropping out of a single gene will act as the breaking of a link in a chain.

The lack of one gene may result in the inactivity of a series of others or rather, we should say that at different moments the development may proceed in one of two different directions. And the coöperation or otherwise of a gene may decide in which direction. In such a case the resulting organism may in one case come under the influence of a set of developmental factors, both genes and non-hereditable influences, which is different in one or several ways from the alternate set. We think that this is the way, in which we must look for the ultimate biomechanical explanation of the difference between the sexes.

The things which are responsible in one sex for a series of reactions, which we do not see in the other sex, must be nevertheless common to both sexes. In animals, like pheasants, where the sexes differ considerably, the result of a cross between two species is the same for reciprocal crossmatings, even in later generations. The characters of the male of a certain species will appear in the first or second generation of the cross, even if a female of that species was used, and vice versa. We need not at all accept the conclusion that, as the ultimate origin of the difference in sex is due to the presence in the female of one gene, absent from the male, this same gene therefore directly causes the visible colour-difference and all the other differences, that the same gene "determines" all these things. At an early stage of the development of the chick the presence or otherwise of this gene may make it respond differently to the action of other genes, will even make that in one a case a series of factors, both genetic and non-genetic may influence the further development and in the other case another series.

If we know from the results of crossing-experiments, that two animals or plants are unlike because of the presence or otherwise of a gene, we need not conclude that the genes themselves are capable of changing one into the other. In our example the sugar-corn and the starch-corn, it may be that the thing which finally changes the sugar into starch is present in both kinds, but that the kind of sugar of the two sorts, in the ripening seeds is not of the same nature. It may be that the sugar in sugar-corn can not be converted by the same agent which causes the conversion of the starch-corn's sugar. The difference between the sugars in this case may be due to a presence or absence of any thing, necessary in the longs series of processes leading ultimately to the production of just this kind of sugar which is capable of being converted into starch by a common thing X, which in itself need not be a gene at all.

If the difference between two individuals, of which one has a gene and the other lacks it is caused by this cooperation or non-coöperation of one single ingredient, one single factor in the series of processes, we see that the same result can be reached in several ways. For the production of pigment in hair or feathers, a whole series of developmental factors must coöperate to the ultimate result. And on several occasions animals may be produced differing from normals in the lack of one gene. The gene need not be the same one, but the result may well be the same, namely albinism.

Instances are known in which albino animals or plants of different families were mated, and in which the cross produced coloured-offspring, showing that the absence of pigment was in each family due to the lack of a different gene. Baur produced pigmented young from a mating of two white rabbits, Bateson found similar cases both in fowls and in the Sweet-pea, and we

recently found a case in fieldrats. The instances in which crossing of two strains, which both had a recessive character of the
same aspect produced individuals with a character dominant
over that of both original kinds, can be found everywhere in the
literature of Mendelism. They are very important in that they
show, how new dominant characters can arise.

Miss Douglas in her work with stocks has found more than
one set of such complementary factors. In one case some plants
were glabrous, instead of hairy, because they missed one gene,
and other plants were glabrous because another gene was
lacking, indispensable for hair-formation. A cross between two
glabrous plants of these
different plants produces
hairy off-spring. Vilmorin
found a similar instance in
the pea, where two species
without waxy gloss produ-
ced hybrids which had it.
It is just as inadmissable in
such cases to say, that we
are dealing with *the two* factors which together produce
the dominant character, as
it is to say that one factor
which we know to be indispensable for a certain process in itself "determines"
this process and the re-
sulting character.

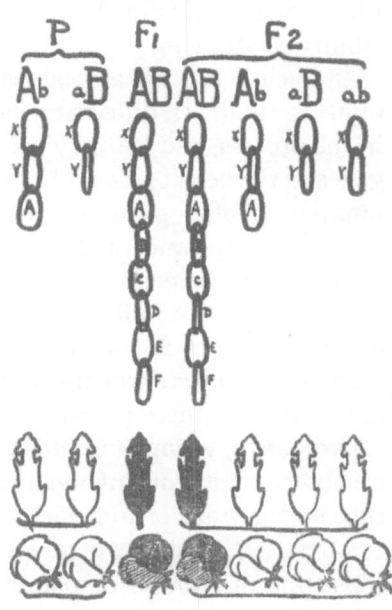

Fig. 6.
The effect of complementary genes. In
this case absence of either A or B breaks
the chain, and makes it impossible for
lower links, genes C, D, E and F to produce their effect upon the end result.

The one or two genes
which we can study,
because they are indispensable for the final result,
may be likened to links in
an iron chain. (Fig. 6.) Any link in the chain may fancy itself to
be *the* link which holds up a weight, in reality all the links are

necessary. If we can only see one link and the weight which hangs from the chain, we may feel sure that this link is in itself responsible for holding up the weight, and if we are fortunate enough to see two or three we may think that a coöperation of these two or three is necessary and sufficient. The difficulty in our analysis comes in because of the circumstance, that as long as a gene is actively doing its work we cannot study it as such. We can on the other hand, not study a gene apart from an organism carrying it, we can only compare the development of two organisms, of which one had the gene in its germ, whereas it was lacking from that of the other. So long as a group of plants or animals are all pure for a given gene, there is no possibility of studying it.

In such a process as pigment-formation, we know in some animals about ten different genes which all play their own rôle in the process, and we may study the effect of each of these genes on the development of animals which have different combinations of other genes. But we have no idea of the number of genes actually concerned with the production of pigment. In the first place we cannot study any genes which are not absent from at least one individual, and in the second place we cannot know in how far other genes which we have actually been studying in their effect upon things other than pigmentation, may influence these processes.

Ordinarily, we must see that variation, the diversity, which we observe in a group of organisms may be either due to a difference in genotype, or to a variation in the environment, or to a combination of the two causes of variability. A population, consisting of animals or plants which have either one or the other of two genotypes may vary continuously, because of the fact that both bio-types are modified in both directions by the environment. But if we are dealing with a population in which all the individuals have the same combination of genes, the same genotype, all variation in this population must be due to the action of the environment exclusively And in such a population we can see if evolution of some kind is still possible.

Through the work of Mendel's followers we now know that a redistribution of genes is ordinarily only possible at the formation of gametes. We must imagine all the cells of a plant or animal to contain the same set of genes, with the exception of the germ-cells, and we have good reason to believe that every cell of an individual is pure for the same genes and impure for the same other ones.

Therefore, if we have a clone, a group of plants or animals, which have all been derived from one single zygote through a sexual process, division, budding, such a group can ordinarily vary only in one way, namely through the action of non-inherited developmental factors. Is the variation which we observe within a clone, of such a nature as to make it possible, that within such a group new species could arise?

If we observe the variation of such common plants as the hawthorn or Taxus baccata, or the holly in hedges or in woods, we see that it is very large indeed. If we should be asked off-hand to give our opinion about the proportion of this total variation which we think is due to difference in genotype and to a varying influence of the environment, we might easily be led to believe that the fluctuating continuous variability observed was in the main due to differences in the condition under which the trees had grown, differences in soil, in shade, in humidity.

Now, in these three plants, clones are propagated by nurserymen. And it is easy in any important nursery to compare the variability in an ordinary mixed lot of Taxus baccata seedlings to that within a row of cuttings or grafts made out of one single individual. I have more than once had occasion to see several rows of this tree, of which each row was a different clone. And if we compare the variability of such a clone with that of a mixed lot, we are struck by the fact that it is so very insignificant. All the trees of the clone have exactly the same habitus. Small things, such as a somewhat concical form, or a somewhat more yellowish hue of the branches, or a slightly darker tinge of the leaves are faithfully reproduced in all the individuals of the clone. The final height to which a four or five year old tree grows is obviously the result of so many developmental factors, influencing the relative strength of terminal and lateral buds, influencing the time at which growth starts, the length and number of internodes, that it would not be surprising if this character, the absolute height of the topmost point, were influenced for a great part by external circumstances. But the fact is, that in a row of young Taxus trees, or Apples, or *Thuyas*, which constitute one clone, the tips of all the trees are almost exactly at the same level. The trees of one clone are so much alike, that if they are not mutilated or marked in any way, it is next to impossible to find again an individual tree at the second inspection of a row.

Now we believed at first, that this absence of variation in clones was apparent, and greatly flattered by the fact that these clones were all commercial-named sorts, which had been selected because of one or more marked characteristics, so that the contrast between the clones threw into relief the similarity between the members of each clone.

Therefore it was important to find instances where a great many colones can be compared, each out out of one individual, without any selection on external characters. It is the practice of the most progressive firms that produce sugar-beet seed to multiply their seed-bearing beets asexually. The firm of Kühn & Co in Naarden, Holland, do this consistently. Every selected root is sprouted in a green-house, and ten rooted sprouts are separately potted up, and later planted out in a row. In the fields of this firm, every summer we can compare hundreds of different clones, all grown with the same care in the same field. The first thing which strikes us if we inspect such a field, is the remarkable similarity between the ten members of each clone. And on closer inspection we see, that even characteristics which in an ordinary culture would assuredly have been looked upon as the consequence of accidents, prove to be due to the genotype of the plant, as all the ten individuals of the clone present them. If one plant has a slightly wilted appearance, with the tips of the leaves hanging down limply, and the tips of the branches hanging down, we would pass it, thinking that some accident had befallen it. But if in such a field of clones we notice a plant in this condition, we see that all ten of the same number have this peculiar appearance. We see a plant whose main stalk stands out of the vertical. We might easily think that a mole had passed underneath the plant, if we did not see all ten plants of the number stand askew in the same way. The ten plants of a clone are either all free from fasciations, or they all tend to be fasciated in the same degree. The degree of development of the main stalk, the moment of shooting from the rozette, is always the same for the members of one clone. Some clones will remain rozettes all summer, other will start sending up a stalk very late, in short, all stages of development are present in the field, always with this restriction, that the individuals of one number are alike. There is so much diversity between clones, that it is always possible to see where one clone ends and the other begins, by observing the general habitus, which is made up of small, inobtrusive diffeences in colour, in branching, in shape. This variation in the field is what goes under the name of individual variation, fluctuating variation. And here we must once more emphasize, that such individual variations are more commonly caused by differences in the genotype than by differences in the circumstances under which the individuals grow up.

After one has observed the nature of the variability within clones, and of variabil ty in mixtures, as due to differences in genotype, one begins to be able to distinguish between them. In other words, one develops this "wonderful intuïtion of the plant breeder" of which the journalists talk so much. It is indeed very important for anyone who intends to produce new plants to develop this faculty of distinguishing at a mere inspection between fluctuating variability caused by genotypic diversity and that caused by the environment. And the only way to develop this faculty is by observing very many pure clones. In plant-breeding it is very important to distinguish be-

tween the two kinds of variation, because variation as induced
by geno-variation is something you can work with, it furnishes
the material for selection, and variation as induced by the
environment is irrelevant from the plant-breeder's view-point.

The variability in the shape of the leaves of the ordinary dandelion, Taraxacum officinalis is enormous. This is surprising if
we take into account the fact that the seed in this plant is produced parthenogenetically or apogamously. No good pollen
seems to have been found in the dandelion. It seems very
likely, however, that occasionally fertilized seeds are produced,
and that new types result from such occurrences. On the other
hand, a very geat part of the diversity in shape of leaves and of
the rozettes, must be due to variability of the non-genetic factors. This became apparent in some selection experiments. One
of us harvested seed from six plants in a garden in Berkeley after cutting-off the upper part of the bud with all the stigmas.
More than half the buds so treated produced an abundance of
normal seed. Two typical large leaves of each plant were harvested at the same time, and their shape was recorded by
making actual blue-prints directly from them. (Fig. 7).

Four sets of plants were raised to maturity. In each of them
there was considerable variability in shape of the leaf. From
each family two extreme plants were chosen. Seed from these
two were harvested, and two leaves of each were blueprinted.
Next season a series of from twenty to twenty-five plants from
each lot were grown in pots under glass in the experimental
gardens of the firm of Vilmorin in Verrières, France. In every
case, the progeny of two very different sister-plants proved to
be identical. Reproductions of the blue-prints will be clearer
than any description. As can be seen from the two series published, the effect of the selection was zero. The plants with the
most deeply serrated leaves in the third year happened to occur
in the progeny of the least serrated mother-plant.

Within one clone we are dealing exclusively with variability
induced by the environment, if we exclude for the present
cases of vegetative segregation in heterozygous individuals. As

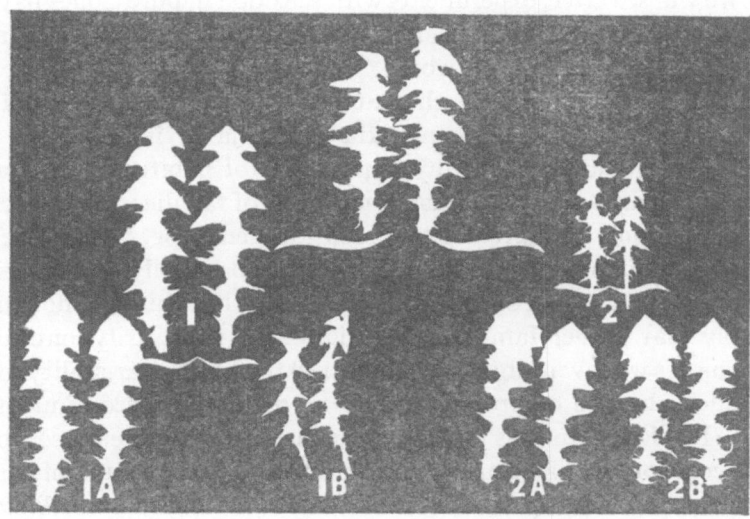

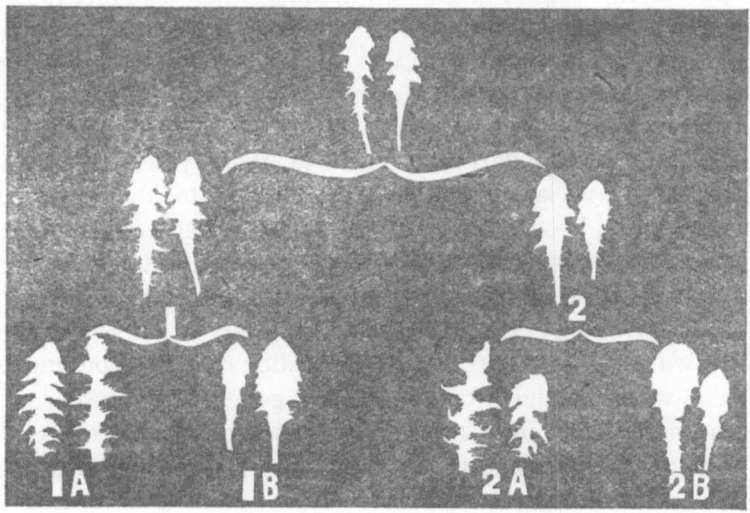

Fig. 7.
The effect of selection on leaf-shape in the dandelion. Nos. 1 and 2 represent daughters of the plant in the upper row, 1A and 1B are two extreme daughters f 1, and 2A and 2B are two extremes found in the progeny of 2.

we will see later, experiments with selection in pure clones have always given negative results, and the apparent exceptions are founded upon a play of words. The negative results of selection in clones show plainly that at least one of the three main theories of evolution, namely that of Lamarck, is untenable.

We will later show, why populations of habitually autogamous plants tend to consist of mixtures of families, which each consist only of individuals of identical genotype. Within each of these families selection has proved to be wholly ineffectual. We think it is not superfluous to repeat, that this does not imply that in such families the variability is necessarily continuous. As we saw in the case of de Vries' teazel, the variability induced by the environment may be distinctly discontinuous.

Inasfar as in this book we are concerned with the process of evolution, we can simplify the question as to the causes of variation, by leaving "modification" out of account, and asking: How do individuals originate which have new combinations of genes, a new genotype?

If we put the question in this way, it becomes clear, that the answer cannot be selection. Selection is of no effect in a population without geno-variability, and therefore it can obviously not produce geno-variation. The idea, that by selection small fluctuating variations can be accumulated into inheritable variations, has been proved untrue. Even these small, fluctuating, continuous variations have been shown to be of two distinct kinds, the result of environmental action and the result of geno-variation.

Two possible causes of geno-variation remain, spontaneous geno-variation and re-combination of genes after crossing, We know that crossing is a cause of geno-variation. It is possible that outside crossing in the widest sense, geno-variation can take place spontaneously? And if so, in how far can we discover the causes of such spontaneous geno-variation?

CROSSING.

AN examination of the causes for variability, for geno-variation, shows that there are two out-standing ones, spontaneous geno-variation or mutation, and the recombination of genes through crossing in the widest sense. We will examine mutation as a cause for that geno-variation which can furnish the material for evolution, and see that, as there is a good reason to assume that mutation ever consists of anything but the spontaneous loss of a gene, it cannot be considered as of any real general importance for the evolution of new species. This brings us to the question whether crossing, re-combination of genes made possible by heterozygosis, is in itself sufficient to cause the genotypix variation which is necessary for evolution.

In the chapter on Heredity we have given the reasons for our assumption that in those cases, where the difference between two alternative characters is due to one gene more or less, dominance means presence, and recessiveness means absence. And here a difficulty comes in. If we see individuals which have new characters, dominant over the corresponding characters of the patent stock, we must admit that this means that these individuals possess genes which were lacking in the parent species. Now there are several instances where such new characters are not found in any other organism, or at least not in closely related organisms. Should we not admit that the corresponding genes have been acquired spontaneously, that their origin has been a special creation? Davenport, among others, has answered this question in the affirmative. As in some tame chickens there exist genes which are absent from the genotype of *Gallus bankiva*, the alleged progenitor of all domestic chickens, positive mutation, the spontaneous acquisition of new genes must, according to Davenport have been a factor in the evolution of these breeds.

So long as we held the view that the inherited things, the genes, were determinants for characters, bound up with these characters in an unvarying relationship, there was indeed no other way out of the difficulty. We know, that those curiously looking chickens which have the feathers reversed and curving outward, have a gene more than normally feathered. If we think of this gene as a determinant of the peculiarity, if we believe it to be in itself responsible for the character, we will have to admit that it has come from nowhere, as we cannot find records of wild chickens with such a curled plumage. We know that polydactily is dominant. If we hold the view, that the gene which in polydactylous chickens determines the aberraton is a real determinant for this character in Weismann's sense, we will have to assume that it has been created de novo, as all the wild gallidae have only four toes. We find that black colour is dominant to the colour of Gallus bankiva. But there are no wild black species of Gallus. Also we do not know buff wild chickens, or bare-necked ones, or chickens with top-knots. So long as we look upon the genes, which determine the difference between animals with all these new characters and their relatives without them, as upon the exclusive cause of these novel adornments, we will have to be content with the idea that they must have been acquired from nowhere, that a special creation was responsible for the origin of each of them.

From the moment we take a biomechanical view of the facts of Genetics, the difficulty disappears. The characters of an individual are determined by the way in which it developed, and the way in which an organism develops is determined by a host of factors, of which some are material and constituents of the protoplasm of all the cells, the inheritable factors, genes, whereas others are of a different nature, together constituting the environment.

A given gene may or may not influence the development, and this the characters of an organism in whose cells it occurs.

But in a given combination of other genes, and other non-genetic developmental factors, its influence will always be the same specific influence.

One and the same gene may have no effect at all upon the development of an organism with a set of genes X in the environment A, whereas it will exert an appreciable influence upon the development of an other organism with a set of genes Y in the same environment A, or even upon the development of the same organism with a genotype X in another environment B.

Take the case of Primula sinensis. There is a gene which makes the difference between certain white and red-flowered plants. In a sufficiently low temperature the presence of this gene in the make-up of an otherwise white-flowered plant, will make the flowers red. In a high temperature it will not do this, it has no influence upon flower-colour. In that high temperature however, another gene may determine red colour. Its influence upon the development is manifestly different from that of the first-named. Neither of these two genes should be called a determinant for red. They determine red colour only in certain special circumstances, in coöperation with a whole set of other circumstances, other factors in the development. There is a dominant yellow in mice, that is to say, certain yellow mice have that colour because of the fact, that they possess a gene which in ordinary conditions of keeping laboratory mice, has such an effect upon the development of the animals that their preponderant colour is yellow. There are no wild mice having a yellow colour. But it may not be said, that there are no wild mice which possess this gene, which will turn the ordinary laboratory mice yellow. Mus wagneri for all we know it, may possess this gene, or Mus sylvaticus, and nothing forces us to assume, that these mice should be yellow if they carried it. Their remaining set of genes is different from that of Mus musculus, and a gene which in ordinary circumstances turns mice with the combinations of genes of musculus yellow may not have that effect in wagneri or sylvaticus. Again, albinos may carry the gene unseen, and they may have derived it from the species whose crossing into musculus gave rise to them.

Two organisms may have the same character, and yet their hybrid off-spring may have an altogether different character, dominant to the corresponding one of both parents.

A definite character of an organism, any quality, results from the coöperation of numerous factors. Some of these factors are quite indispensable for the final result. If one such is lacking, the character is not developed. If another one is lacking the result is the same. One rat may be albino because in the chain of factors necessary for pigmentation one link is missing. Another rat may be albino for quite a different reason, because of the lack of some other factor. The result will be, that the children of two such different albinos will be coloured.

The first cases recorded of such an unexpected production of pigmented organisms from the cross of two whites with the same recessive character, were found by Bateson and his pupils. There is first of all the case of the two white sweet-peas in Emily Henderson, both whites, differing chiefly in shape of the pollen. The hybrids where purple. On analysis it was proved that one was white because it lacked A, and the other was white because it was lacking in a gene B. Together however, the gametes constituting the hybrid zygotes contained both A and B, and as all the other genes necessary for pigmentation were common to both parents the hybrids were coloured. Such coloured hybrids will produce as many gametes with as without A, and the same holds true for B, for that reason they will, if test-mated to individuals lacking both A and B, produce three times as many white as coloured off-spring. But if they are mated black to either parent, the proportion of coloured and white will be equal. For such hybrids differ from each of the parents in one gene only, namely in the gene inherited from the other parent. Coloured hybrids from white parents have only one gene more than each parent. If we mate them back with the parent having A but lacking B, we are not concerned with A, and only with B, as A is common to parent and hybrid. In other words, in such cases of testmating coloured hybrids, we find mono-factorial segregation. If we are growing Emily Henderson sweet-peas, and find an unexpected colour-ed plant, this may be a hybrid from pollen of a coloured plant, but equally as well from pollen of another white. Mated back into our Emily Henderson strain it will probably show to have one gene more, here responsible for pigmentation. If we are lucky enough to lay our hands upon the original F, hybrid, we can in the case of the sweet-pea test it by self-fertilization. In this case the ratio between its coloured and white-flowered off-spring will be 9 : 7 in the case its unknown parent was a white, and 3 : 1 if it was a coloured.

But it is evident, that if we did not discover coloured-plants in our strain till the next generation, we might find that a coloured plant in a white strain gave a 3 : 1 ratio of coloured to white off-spring, even if its other unknown grand-parent were a white.

The most important fact we think, is this, that if we discover a dominant novelty, we can ordinarily not decide whether it owes this new dominant character to the spontaneous crea-tion of a new gene, or to a combination of genes which each do not determine the novelty. The third possibility, namely, the acquisition of the gene from a form in which it determined the same character, can as a rule be excluded by inspection of related forms.

This case of the two white sweet-peas giving a dominant coloured-form was soon followed by a great number of instan-ces in which two parents, having a similar recessive character produced hybrids with a corresponding dominant character. In most instances the recessive character was not quite the same in both parents, such as it was in the sweet-peas. The two white chickens in Bateson's work which gave coloured young had a different down-colour. The white Viennese rabbit, which Baur found to give pigmented young with pink-eyed albinos has itself blue eyes. The two yellow rats which originated in England a few years ago, and which, when crossed, give wild-coloured young differ in colour. The two Stizolobium forms, from which Bellings obtained hybrids with burning hairs, differed in the degree of hairiness.

Recently we observed a case which is quite analogous to Bateson's case in the sweet-pea. Mr Spanjaard, manager of the Ketangoengan sugar factory on Java, sent us some young field-rats of different colours. Among them, was an albino female, who produced numerous wild-coloured young from matings to a male field-rat from Sumatra. These young, when mated together, produced ten albino young and thirteen pigmented. Only one litter was raised from the original female and an albino son. It contained four albinos and two with pigmented eyes, but was destroyed by the mother before it was four days old. Two albino F2 were left to continue the work with. They produced three litters, one of six, the second of five, and the third of three. Not one albino was amongst them, all being pigmented. The only possible conclusion is that the original albino female must have been lacking in two different genes, each indispensable for pigment-formation. The origin of the albino female is unknown. But everything points to it that the group of rats in which she was found were the descendants of a cross. For with the exception of this albino, two black-eyed whites or creams, and two very pale greys were caught, and the inheritance of the waltzing character turning up in the F_2 of the albino cross is as complicated as that of albinism. Waltzing in house-rats we found to be recessive.

One pair of the coloured young from two albino parents have up till now produced 10 coloured and 6 albinos. A second female had 5 coloured and 6 albinos. The ratio of coloured to albino in the first pair is sufficiently close to a 9 : 7 one, two make the existence of two genes probables.

In numerous instances new, or comparatively new, characters must be produced by crossing by combining genes, which up to the moment of the cross were present only in different individuals. The presence of the burning hairs on the pods of Belling's hybrids between the Lyon bean and velvet bean is a very good example. Nearly every cross between animals and plants, which are not too closely related produces new characters as its result.

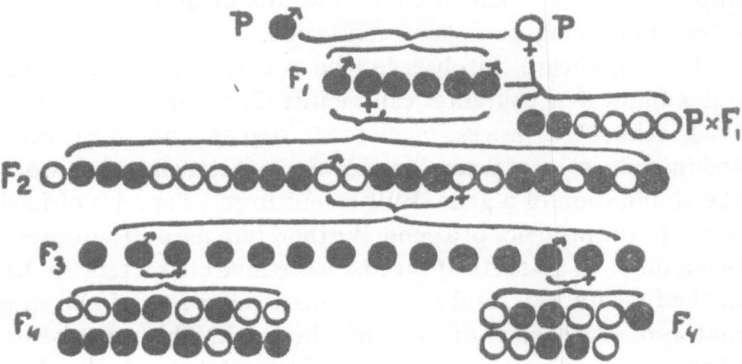

Fig. 8.
Inheritance of albinism in a family of Malayan field-rats.

We do not think it probable, that new breeds of animals or plants, other than those which are habitually self-fertilized, are directly derived from hybrid individuals. Even in the history of the tame species of fowls, not many instances are known, in which a new breed was deliberately produced by crossing, out of the variable off-spring of hybrid birds. The common way in which new breeds of fowls are made, is by breeding a new character into an old species. Species with a colour, new for the group, are often produced, by breeding an animal of the desired colour with good typical representatives of the species, and by continuing to breed the hybrid back to the old species until the desired result is obtained. It is especially easy to introduce a gene, not heretofore present, as the desired character which in this species greatly depends upon the presence of the gene, is not lost sight of. In our size-inheritance work with mice, we find it extremely easy to retain any one gene present in one species during a four or five times repeated crossing back, "grading up" to a second species. By breeding back to an albino strain, without more selection than retaining any pigmented animal, the final outcome of the work will be coloured animals, which differ from the albinos only in that one factor which all albinos lack. In this way a good insight is obtained into the genetic constitution of the albino strain, so far as colour-modifying genes are concerned.

It is significant to observe, the relative ease with which a dominant new character can be introduced into a species of fowls, makes this process much more frequent than the reverse, the introduction of a new recessive character, in other words the elimination of a gene. Buff colour in most species of fowls is due to the presence of a gene. We therefore see buff sub-breeds being made in all sorts of breeds. Recessive characters are lost in the first hybrids, and this fact often throws would-be originators of new species of fowls off the track. This must be one of the main reasons why there is only one colour of Silky fowl; to a dozen colours of booted bantams, notwithstanding the popularity of the silky breed. And that there is only one colour of silky fantail-pigeon, to a dozen or more colours of normally feathered ones, notwithstanding the striking appearance of the white silkies.

As a rule, fanciers of fowls are content with a new species, long before it is wholly up to the standard of excellence of the rest of the species of the group, and some traces of the cross will often remain.

If the view of some authors on the origin of bantam fowls were correct, that bantams could be made out of any of the breeds of fowls by selection of small "mutants," it would appear strange, that they hardly ever are good copies in miniature of the breed of which they are believed to be descended. In answer to questions directed to breeders of bantams, Poarl learned that bantams are bred from hybrids between the large breed and some bantam. From the history of one of the newer bantams we know that in practice the whole process is more a "grading up" than a straight selection in the second generation of the cross.

According to Bateson, there is only one striking dominant

character in Primula˜sinensis, which could be thought to be due to progressive mutation. This is the absence of the yellow eye as seen in "Queen Alexandra" Gregory found that the hybrids between such plants and normals, when self-fertilized, gave a mono-hybrid 3 : 1 ratio. In his books on "Problems of genetics" Bateson discusses this point. On page 92 he writes: "There is no real doubt that it came into existence by the definite addition of a new factor, for if it was simply a case of the appearance of a new character made by combination of two previously existing complimentary factors we should expect that, when Queen Alexandra was self-fertilized a 9 : 7 ratio would be a fairly common result, which is not in practice found".

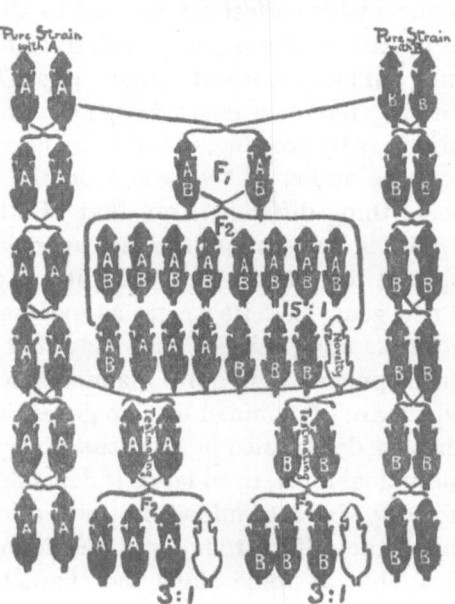

Now here we must again distinguish between the original hybrid, and later individuals with the same character. For if the dominant character in question really results from a development to

Fig. 9.
Diagram illustrating the origin of a "double recessive" novelty, and the fact that it closely simulates mutation, and cannot readily be distinguished from it even by test matings.

which two genes contribute, which are not found together in the genotype of other species, hybrids with Queen Alexandra as one parent, and exhibiting the new character, would not in self-fertilization produce a 9 : 7 ratio, but a 3 : 1 one, only one gene being concerned. Only those plants

5

can be expected to produce a 9 : 7 ratio, which exhibit the character for the first time, and those of their off-spring which happen to be heterozygous for both genes. We will see in a later chapter how small the chances are, that a heterozygosis for two genes will persist for a number of generations.

If we see that polydactylous chickens, or black-skinned ones, or animals with the feathers recurved, have only one gene more than the normals with which we cross them, we may assume with perfect safety, that in all these cases, such genes, as are new to the species, are derived from other species, which may not have shown the peculiarity. The main point to remember is, that new characters, new dominant characters, can originate by crossing, in all those instances in which the simultaneous action of two genes on the development results in some thing different from that of either alone.

The circumstantial evidence upon which Bateson would feel obliged to concede the possibility of a spontaneous origin of a new gene, loses its weight as soon as we decide to look upon genes as upon things which may, but need not, influence the development. It is quite inadmissable to speak of characters which are determined by two genes, in contrast to characters that are determined by one gene. Every single character of any individual must, in so far as it depends upon the genotype, the heredity, be determined by quite a long list of genes, acting on all the different stages of development, and so influencing this, that at some stage the character, the quality studied results. We know the genes only through the difference their presence can make in the development of an organism, as contrasted to the development of another organism, which lacks them. In some instances we know one gene in this way, in other cases we may know to or three genes, which, when coöperating to the development of a line of organisms affect the final result on the same point. When Baur and one of us came out in favour of the so-called presence and absence theory, Plate criticized us. He had the impression that we believed that a definite quality could result from the absence of a gene, and

he offered a "Grundfactor-Supplement Theorie". If we substitute "residual genotype" for Plate's "base-factor" his hypothesis is in full accord with ours. If two organisms have an
identical set of genes with the exception that one has an additional gene, we have no great objection to calling this gene a
"supplement." But no matter how we call things, we can have
only a faint idea of the complexity of the whole development,
up to the point where the presence of the gene under discussion makes itself felt. What we should do, is to try to study the
way, in which a gene does affect the development, and thus,
in the language of the Weismann-Mendelians "determines its
unit-character".

The cases in which we happen to know three or six genes
which all influence the same quality of an organism, may
seem more complicated than the cases, in which we know of
only one. They are certainly more difficult to analyse. If we
know from Nilsson Ehlès work, that there are at least three
genes in wheat which tend to deepen the red colour, whereas
we know of only one gene in wheat which tends to make the
colour black, this does not mean, that the processes which
ultimately lead to the production of red grain are any more
complex than those which produce black grain.

Many of the Geneticians still hold to the Weismannian idea
of determinants for characters, and to them the genes are
transmitted things which are in themselves responsible in some
way for definite characters. They think with Castle that it
would be theoretically possible to "analyse the characters of
an organism into the component units." This conception of
characters of organisms as things in themselves, and of organisms as mosaïcs of charcters has unavoidably grown out of
the circumstance that the "rediscovery" of Mendel's work by
Correns and Tschermack came just at the right time to dove-
tail in with de Vries' revival of Darwin's pangene-theory.
Further, just when de Vries had set forth his conception of inheritance as the transmission of numerous different pangenes,
each responsible for a character, and his belief that evolution

was essentially the spontaneous creation of new pangenes, the cytologists had made remarkable progress in the technique of studying chromosomes.

Most of the genes studied by Mendel in the pea, and by the first Geneticians who verified Mendel's work with all kinds of animals and plants, were of such a nature that they influenced the development considerably, and at different points. The material was carefully selected with this point in view. And it was easy to believe, that the mutual independence of the characters studied, depended directly upon a corresponding mutual independence of its genes, and further, that a certain character would be present, when its determinant was present, and absent when it was absent.

And after the idea had once become established that each character had its own representant in the germ, its own factor, the cases, which were discovered a little later, in which we saw two or three seperately transmitted genes influence one identical quality, were looked upon as complications.

Even nowadays, we read about characters which have a multiple representation in the germ, and it is evident that some authors consider such cases to be exceptions from a rule, that every character of an organism usually has its own single determiner.

And for this reason it is not strange, that some authors have even tried to explain those cases, in which more than one gene were found to influence a certain quality, by the assumption that here a single gene was somehow split up into component ones, that such sets of genes had more in common than their effect on the development, that they were fundamentally of the same nature. We think, that the facts as we know them, do not at all force us to the assumption, that in those cases, in which we know more than one gene influencing the same character, these genes must be fundamentally analogous. If the only thing we know about three genes is, that they all tend to make grain-colour darker, or that they tend to make a plant grow taller, it is easy to be led to believe, that they must be in some way essentially alike, that they may be even part of one pre-existing whole. But as soon as we know a little more about such a set of genes, and especially from a biomechanic point of view, that is, if we begin to see in which way, physiologically, each tends to bring about its effect on the given quality, we see that fundamentally they may have no more in common, than they have with genes which influence quite different qualities. In the mouse we know a good many genes which influence the colour of the eye. Some of these tend to make the colour lighter, and some tend to make it darker. These genes have all been studied in their effect upon the coat-colour, and as they all materially affect this colour, it has been possible to study them, to distinguish them, one from the other. Only incidentally it has been found, that they modify eye-colour.

If someone had set himself to study eye-colour in mice, without attention to coat-colours, he would have had enormous difficulties. He might have found the approximate number of the genes concerned in the variation of eye-colour in his material, but without recourse to a study of the effect of his genes upon coat-colour, he would never have been able to make a complete analysis. Very probably he would have concluded that these genes must be fundamentally analogues.

Philippe de Vilmorin has studied quite a number of different genes in the pea, which influence the degree of coherence of the seeds in one pod. The influence of different genes proved to be very different. In the first place, one gene was found to be present in most peas, whose presence so modified the texture of the seed-coat that even in the most favourable combination of others, adjacent grains were almost always wholly free from each other. In the absence of this gene, coherence of adjacent seeds was a common phenomenon, but the degree of sticking together was greatly influenced by other genes, in both directions, at least by six.

The genes which in certain types produce the difference between un-pigmented, yellow or green, and pigmented seeds, and the difference between plants with rose and purple flowers gave the seed-coat a lesser tendency to stick. And it was found that the waxy gloss, common to most peas materially hindered the coherence of the seeds. Two genes were found which are both indispensable for a formation of this waxy gloss. Plants lacking in either or both of these two genes have the seeds and all other organs devoid of wax, and the seeds in such plants hang together in the ripe pods with a much greater frequency than in common peas. The gene which is present in most plants with hard stiff pods, but lacking from those with soft, clinging, shriveling pods has a very great influence upon the character. No matter what the further genotype may be, plants which lack this gene, and which consequently have soft pods, will hardly ever be found to have coherent seeds. In the shriveling of the pod the seeds are broken apart. In a similar way the character is affected by the genes which affect the relative size of seeds and pods, for those seeds which are not in close contact when ripening will never hang together. And lastly, the gene which round seeds have more than wrinkled, will tend by its presence to further the coherence. In pods which contain both wrinkled and round seeds, two adjacent rounds will hang together with more tenacity than two wrinkled ones, or one round and one wrinkled.

If anyone would want to study the coherence of seeds in the pea as a separate character, and without reference to the shape of the pods, the nature of the epiderm, the presence or otherwise of wax and of pigment, he should be considered very lucky if he could merely estimate the number of genes influencing the character. He would be unable to distinguish the genes, and most assuredly he would tend to the assumption that these genes must be fundamentally analogus in some way.

Lang has proposed the name "polymery" for cases in which a number of genes influence one character. The objection to the use of this term, and similar ones, is obviously, that their use tends to make appear exceptional what is essentially common. The use of the term "polymery" for those instances, in which we know several genes which influence one character, makes it appear as if ordinary characters were determined each by a single gene, and it will help to strengthen the belief in "unit-characters" as contrasted to characters which have a "multiple representation" in the germ. The difference in quality between two organisms may be essentially due to presence or absence of one and the same gene, but in each of the two any quality as such has to be conceived as a result of the development up to the moment, at which we can notice the quality, under the influence of a host of developmental factors, genes and environmental factors.

If the study of Genetics had come sooner under the influence of Biomechanics, if for instance Roux instead of Tschermack had "rediscovered" Mendel's work, the theoretical side of this new science would not have come under the influence of the Weismann—de Vries conceptions about "determinants".

If from the outset we had seen, that "Mendelism" afforded a means of studying the inherited development-factors as postulated by Wilhelm Roux, there would have been no difficulty and no confusion, it would have been clear that some of these factors might influence the same process, and that others might influence different developmental processes.

The object of this digression was to show that, if we see new dominant qualities produced by crossing, we must not assume that such new qualities, as they are in this case "determined" by at least two genes, differ from ordinary characters which are determined by only one. Each gene, new for the strain into which it enters, "determines" a character only in collaboration with a long series of other genes, already present.

In some instances, the fact that a new gene enters into the composition of animals or plants of a given family, compels genes which were already present, to participate, where they were inactive before. In fact, this must be the common way in which new dominant characters originate. Bateson has proposed to call genes whose coöperation produces an effect, different from that of the action of each alone, complimentary factors. Tschermack uses the name "Kryptomere" for genes

which are shown to be present, without exerting any influence upon the development of the organism in which they
occur.

The danger in the use of such special terms is that they make
appear exceptional what is usual.

Crossing is practically the only cause of the production of
new dominant characters in a group of organisms. But it is
evident, that the new character as such need not be "introduced" from a stock "carrying" it. Usually a new dominant
character, resulting from crossing must be looked upon as due
to a difference in the development from that in the common
type, caused by the presence in the zygote of one or more genes
not commonly present. Such "new" genes may so affect the
development in coöperation with all the other developmental
factors, that a new character results. The cross need not be a
wide one. Within one species with a sufficiently high potential
variability, combinations of individuals may be possible which
give off-spring with unexpected new characters. Dominant varieties which can be made into domestic species, may originate
in this way, and such an instance may easily be mistaken for
positive mutation, for the spontaneous acquisition of a gene.
(Drosophila).

New recessive characters may result from cross-breeding, and
it is evident, that this process looks so much like loss-mutation,
that it is only with the greatest difficulty, that it can be distinguished from it. Lotsy has accepted our view that almost
all the instances of loss-mutation recorded must have been
cases of the production of new forms, new characters, through
re-combination of genes. We would not subscribe to his statement that mutation is a myth, because it appears difficult to
account for such mutations as have been seen to take place in
"pure lines" by both Johannsen and Nillsson Ehle, without
assuming that here a gene, present in the material was excluded
from some cells. If we remember, that every gene present
in a cell must be multiplied quantitatively and distributed over
millions of descendant cells, we can well imagine the possibil-

ity of a hitch in the process, resulting in the absence of some gene in a few cells.

It must however, be well-nigh impossible to distinguish between a real loss-mutation in animals and the production of a recessive novelty through re-arrangement of genes. And it is possible that, notwithstanding the breeding-tests performed upon the "mutating" individual mice in our colour-inheritance series, the production of the recessive novelties may have been caused by simultaneous absence of two genes in respect to which each grand-parent was pure for one.

In the light of this difficulty, it becomes significant, that hardly any cases of "dominant mutations" are recorded, and none at all in "pure lines. "We think, we can, practically speaking, exclude mutation as a factor in that variation, which can furnish the material for evolution.

A number of very good instances of the origin of recessive novelties, which simulate loss-mutations have been studied by Brainerd. He found numerous cases, in which apparently a good species had spontaneously produced one or more varieties. He found a few Viola affinis plants with black seeds instead of seeds of the normal yellow colour. These plants bred quite true. In every character, other than seed-colour they were typically normal Viola affinis. Another plant found was a normal Viola cucullata with the exception of having dark purple capsules instead of clear green ones. A plant of Viola nephrophylla was found, which carried buff seeds instead of black ones and which bred quite true to the new character. Later Brainerd could show by cross-breeding experiments that these apparent mutations were the result of crossing between species. He was able to reproduce them and produce new similar instances.

Very striking examples of the origin of new characters by cross-breeding are found in a series of experiments with rats of the Mus rattus group, which we have been continuing for a number of years. Some of this work was started by Bonhote, and continued by us, when he left Europe. Bonhote mated white-bellied tree-rats from Egypt with the gray-bellied roof-rats. The hybrids had grey-bellies, but produced some, 25%, white-bellied off-spring. When a few yellow rats had been produced, we took over the work. Most of the rats died in the first attempt to pass them through the French custom office, and we finally received only a few white-bellied rats, of which only two ever bred. One female was successfully mated to a French black Mus rattus male. The hybrids were black, of a colour somewhat different from that of the pure French house-rats. From these hybrids, bred together and bred to one of Bonhote's males, we obtained several new colours. We obtained some yellows, as was to be expected from Bonhote's work, but we also obtained a cinnamon agouti, and three chocolates, several rats of a bright silver colour, like that of a lilac mouse, rats with a bright yellow belly and an agouti coat, animals with a white tip to the tail, and waltzers.

If we add together the young from two F1 sisters, which gave waltzers, we find that they had thirty-eight young. Among these there were as yet no chocolates, and none with yellow bellies; these came in later generations. But among the thirty-eight there were four yellows, four waltzers, one silver, and two with white tail-tips. These numbers show that this was not a case of mutation, but of the production of new characters through recombination. If in the production of every new recessive character, we would have been dealing with a case of mutation, we would have expected a proportion of one in four. If, however, we look to the cross itself for the production of the new varietal characters, we can see how each novelty may have originated where two genes, each present in one species, were both absent from a zygote. In that case we would expect the novel characters to be present in one animal among sixteen. Now the frequency of the different new characters is differently great, but if we calculate the average frequency, we get a proportion of 2.7 to 38, which is sufficiently close to one in sixteen to be significant. If we remember that chocolates and yellow-bellied rats cropped out in the next two generations, it becomes clear that each of the five new recessive characters must have originated as the result of crossing. In our work with the house-rat group we saw the origin of several new recessive characters, and we were in a position to study the behaviour of these new characters in crosses,immediately upon their origin. In no case did we observe a mono-factorial difference between the parents and the first representatives of the new varieties. But we often found quite normal uni-factorial ratios in crossing the new colours with pure-bred un-related rats. For instance, we found that the waltzing character, although it originated by a "double absence" behaved as a simple recessive, if we tested our waltzers of composite origin by mating them to pure house-rats from Holland. And the same held true for yellow colour. Although it obviously originated by the simultaneous absence of two genes, one of which only was present in each of two wild species, the yellow colour behaved as a simple recessive in a cross with wild Javanese house-rats.

From the hybrids between our first waltzing male and a wild-caught house-rat from Holland, we obtained three waltzers among eleven young in two litters, obviously a mono-factorial ratio, showing that the difference between the waltzer and the wild rat was a difference in presence and absence of one gene.

And when we mated hybrids between a yellow male and a Javanese house-rat female back to yellows, we obtained a mono-factorial ratio, five agouti to five yellows, in three litters.

In the last few years we have also been cross-breeding strains of field-rats, belonging to the Mus Rattus group. The common field-rat found throughout the Malayan Archipelago, from Singapore south through Sumatra and Java to Bali is remarkably homogeneous. The rats from south Sumatra, which we are still breeding, have a tail which may be very slightly longer than that of the Javanese rats, but otherwise we have not beeen able to find any constant difference, even in an examination of hundreds of freshly-killed rats. The hybrids between these two rats are in every way similar to both parents. In the

F2 generation however, we obtained waltzers. These are of
course in no way related to the waltzers in the house-rat series.
Both physiologically and genetically these waltzing rats behave
differently. They spin around with the same speed as house-rat
waltzers or waltzing mice, but they are better able to take care
of themselves. They can climb fairly well, whereas the house-rat
waltzers cannot. They can walk around their cage, and eat,
without having to interrupt their meal for seemingly involun-
tary spinning. The inheritance of this character in this series of
rats is not yet worked out. Curiously enough, we find, that
matings of two waltzing rats can produce all normal off-spring,
so that it seems as if we had at least two genetically different
kinds of waltzing field-rats. The analysis in this case is made
very difficult, and may prove to be impossible, by the fact
that waltzing females have never in our experience been known
to raise their young. A foster-mother must be provided for the
few young which are not killed by the mother at birth. A
complete litter of living young from a waltzing female can be
obtained only very rarely.

No evidence for the production of novelties has ever been
found by us in pure families of wild rats, not even in the families which produced the novelties when inter-crossed. In several instances, however, rats of aberrant colour have been
caught wild. In one case we obtained several new colours in
one shipment, among rats all caught in one locality. From the
genetic behaviour of this group it is probable that a cross with
some other species, with the house-rat or the tree-rat or Mus
concolor caused the production of the new colours.

In these few examples from our work with rats, we saw the
origin of waltzing individuals in two separate series. In both
cases waltzing rats originated in the F2 of a cross between species which were very closely related, in the last case pheno-
typically indistinguishable. This origin of waltzing in rats gives
an explanation of the way in which the same variation may
have originated in mice. Most authors who have studied the
anatomical, physiological and psychological peculiarities of

waltzing mice have looked upon these animals as upon a variety of the house-mouse, Mus musculus. Droogleever Fortuyn however, found that the ordinary Japanese waltzing-mice had characters which distinguished them Zoölogically from Mus nusculus, and made it appear to be a variety of the Oriental house-mouse, Mus Wagneri. The short tail and small size of these mice are not due to the genotypic aberration which causes the forced movements. We found that in cross-breeding Japanese waltzers with normal mice, small size, tail-length and waltzing were independent characters, which could be found differently combined in F2 individuals. Recently, we found that the ordinary tame mice of the Orient which we imported from Hong Kong and Japan, had the same size and tail-length as waltzers from those regions. It is very probable, that the cause of the production of the first waltzing mice has been a cross between house-mice of different countries, for instance between Mus musculus and native house-mice of Japan. The same cross may have produced other novelties, such as the pink-eyed coloured forms which are derived from Oriental mice, and are frequently met with among oriental stock.

The de Vries—Weismann speculations have intimately permeated Genetics, and it is not surprising that almost everybody thinks of hybridization only as a way in which existing characters may be recombined. Genes are by the majority of Geneticians still looked upon as determinants for characters, and recombination of genes is thought to be the same thing as recombination of characters.

If, however, we think of genes as of things transmitted through the germ, which may in given circumstances, and in given combinations of other genes influence the development, it becomes clear at once that recombination of genes often must be the cause of new characters, dominant characters as well as recessive ones.

If we mate a black waltzing mouse to a chocolate normal, the young will have a combination of the blackness of the

father and the normal progression of the mother, and among
their children we will meet with all four possible combinations,
with the combinations of colour and behaviour common to the
grand-parents, but also new combinations, black normals and
chocolate waltzers. In this case, where we are dealing with only
two genes, the remaining genotype being identical, and where
the action of each of the two does not interfere with the action
of the other, we have an instance of recombination of genes
which goes parallel with a corresponding recombination of parental characters. But if we mate a yellow-agouti mouse to a
black one, we will obtain two new forms, an agouti in F_1 and
tortoise in F_2, and it will not be possible to speak of these new
forms as of recombinations of parental characters.

Just as the Japanese waltzing mice differ from other domestic mice not only in their most striking character, but also in
several others, the domestic rats, and the domestic guinea-pigs
and rabbits do not only differ from wild species in their most
striking and valued characters. They show traces of their origin by hybridization in certain characters, which are either
new, or which they have in common with different related wild
species.

The ordinary albino, and hooded black tame rats, bred for
laboratory purposes are commonly looked upon as varieties of
the Norway rat, differing from it in a few genes. It is true that
they will hybridize with this wild rat, and that the hybrids are
fertile. But they are nevertheless consistently different from
the Norway rat, unless recently crossed with it. These hooded
rats originated in Japan, and the way in which they have been
produced is obscure. They may have originated from a cross
between Mus norvegicus with its oriental representative or with
some other oriental rat. We have examined a strain of albino
rats bred from wild albino variants of the Norway rat, caught
in the sewers of Paris. These albino rats were true Norway rats
in everything but colour, and they were very different indeed
from the real pure-bred laboratory rat from Japan. Nobody
could mistake them. The laboratory rats, when pure, are very

much smaller than Norvegicus. They have large prominent ears and eyes, a convex nose, harder, sparser, glossier hair, a long thin tail with a soft, rather loose skin. The skin of the tail is liable to break and slip off just as it does in rats of the rattus group. This never happens in norvegicus animals, which can safely be caught and held by the tail. The ears of the Japanese rats are naked. Their eyes are placed at the sides of the head, and the small sunken eyes placed close together on the top of the flat head of Mus norvegicus individuals, is seen in white and spotted rats only if they are descendants of a cross with wild rats.

To one who is accustomed to handle living rats of many different species, the difference in disposition is striking. The quality of the coat is also very different. Norvegicus has fine, close hair and an abundance of wooly under-fur, which makes the coat water-proof.

This difference between the soft, furry coat of the Norway rat and the hairier, glossier, harder coat of the hooded rats becomes very striking if the animals are made to swim. The laboratory rats, if of pure origin, become soaking wet and lose their shape if immersed. Norway rats, including albino "sports" and the yellow wild animals of the strain which were caught in London a few years ago, will dive and come up practically dry, only the outside of the coat being wetted. In fact the ordinary laboratory rats look so much like rats of the Mus rattus group, that it is not surprising, that they have long been regarded as a tame variety of Mus rattus. In cranial characters however, the laboratory rats are somewhat closer to norvegicus than to rattus, although the difference in skull between them and wild Norway rats is striking.

The size, fertility, disposition of the domestic cavies differ as much from those of the different wild species as those same characters in the rats. From analogy it would appear probable, that the variability in the cavy which made possible its development as a domestic animal, resulted from crosses between wild species.

In another chapter, the question whether any considerable group of domestic species can originate from one single domest- icated one, will be more fully discussed.

In garden plants, there exist a number of aberrations from normal colour and normal shape, which frequently recur in the most diverse groups. Partial and complete loss of pigmentation laciniations, doubleness of flowers are characters which are found in varieties of the most diverse cultivated plants, and which are found to breed true in domestic species. Mutation seems at the first view to be the most probable cause of the production of these novelties, but the example of the rats should make us cautious. Is it not possible that such new char- acters have resulted from crossing, from re-combinations of genes such as they were present and absent in the genotype of two or more species or even sub-species?

Erwin Baur's work with snap-dragons furnishes a good many instances of the origin of wholly new characters in second and later generations of species-crosses. In the F_2 of a cross be- tween Antirrhinum majus and A. molle, Baur obtained a great number of novelties. Among these, the greatest number were what the Weismannians would call re-combinations of parental

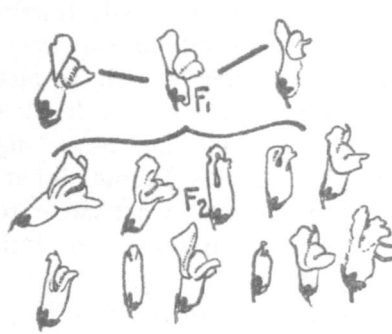

characters, but several plants showed qualities which are wholly new, not only for the two species crossed, but for the whole genus Antirrhi- num. Among other things, plants were produced with peculiar appendages, fringes. Other plants from species hybrids produced branches of wholly female flowers without petals standing in the place of the flowers.

Fig. 10.
Variability in the F2 of an Antir- rhinum cross, after Baur.

The seed-firm of Haage and Schmidt in Erfurt crossed mari- golds, Dimorphocotheca aurantiaca and Calendula pluviatilis.

In the F_2 of this cross between species, the variability was prodigious. All sorts of new colours were produced, such as salmon, lemon yellow, blueish, and some plants had more or less double flowers. Double-flowered strains of different colours have by now been fixed as domestic species. The seed firm of

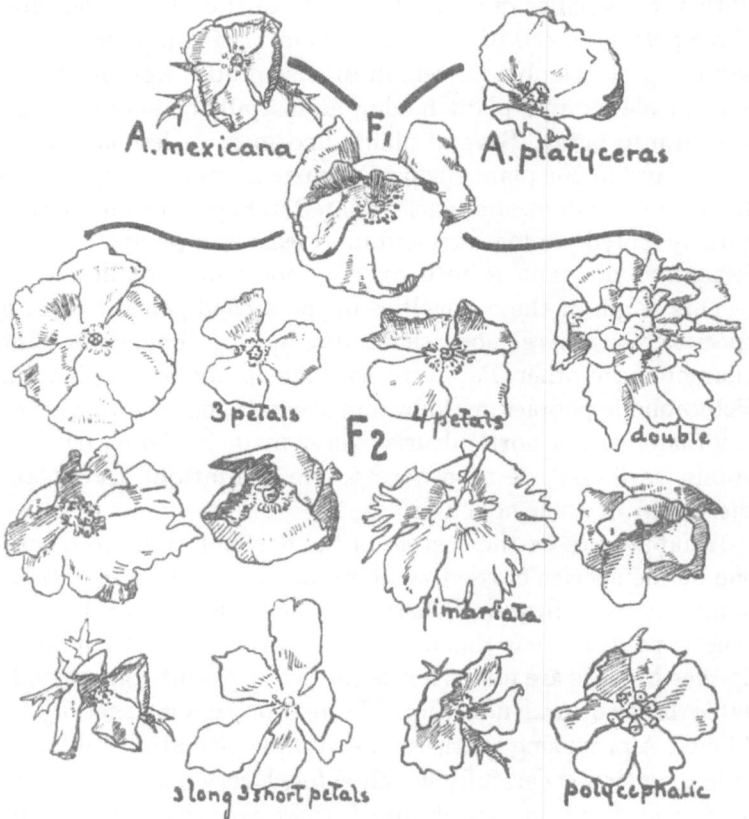

Fig. 11.
The result of a cross between two species of Argemone.

Vilmorin Andrieux and C°. in Verrières, produced hybrids between two species of horned poppy, Argemone mexicana and A. platyceras. Both these species have flowers composed of six petals of equal size, small and yellow in mexicana, and big and

6

white in platyceras. In the second generation the variability
was stupendous. (Fig. 11). A. whole list of novel characters was
found. Some plants had purplish stems, a character unknown
in either parent-species. As to the flowers, a number of F_2
plants had petals of unequal size, three long ones alternating
with three smaller ones. At least two of the plants had only
three petals instead of six. New colours were many, including
salmon pink. Double flowers in diverse grades were found on
several plants, one plant having almost all the stamens converted into petals. Several plants had more or less laciniated
petals, and in one plant the petals where as much fringed, as in
the laciniated domestic species related to Papaver somniferum.
Finally, polycephaly was noted in at least two plants, a transformation of stamens into ovaries, containing ovules.

This origin of these novelties in the second generation of a
species cross in Argemone shows how wholly similar aberrant
characters in other Papaveracea have probably originated.
Polycephalic poppies are known in several species. As we know
how many crosses horticulturists have made in those plants to
obtain variation, we need not assume "mutation" to explain
the origin of such novelties.

It happens to be the custom of the horticulturists to regard
one of the species crossed as more important than the other,
in almost every instance where crossing is resorted to. From a
genetic point of view this practice is on the whole defendable.
Species-hybrids are not as a rule the direct parents of new valuable horticultural novelties. The real process is generally as
follows. A promising species is taken into cultivation, and possible variation is carefully watched for. Every possible cross is
tried, and if hybrids are obtained, these are crossed back into
the species. Variability is the result, and from now on novelties
begin to crop out. These are separated, and variously combined. Newly imported related species are eagerly sought after,
as furnishing material for crosses. From personal observation
of the work of horticulturists, we know that the object of cross-
breeding with a newly imported species does not appear to be

primarily the wish to re-combine valuable characters which the new introduction may possess, with the qualities of the species in cultivation, but almost exclusively to break up the constancy, the purity, the lack of variability of the species on hand. The object of crossing is to produce variability, and the object of producing variability is to obtain the material for selections.

A newly imported wild-daffodil, or a new Gladiolus, or a new primrose, is hailed as an important find by the specialists. And the question which the knowing ones ask, is not whether it has desirable new characters, but whether it is new, that is, whether or not it has already been crossed into the domestic species of daffodils, or roses, or primroses. If it has not been usedbefore, the chances are, that the variability which crossing with it will produce, will cause novelties to appear later on, which.will be real novelties, and not only reproductions of things seen before or re-combinations of qualities of domestic species. In other words new species are sought by horticulturists which will heighten the potential variability of their stock. In some few instances hybrids will be produced, which as a group have some distinctive set of characters, which will set them apart from all the parent-species which went into the combination. In such a case the horticulturists will not name this group of hybrids after one parent group, but they will give a new name to the group, mostly the name of the genus combined with "hybridum".

In most instances however, a group of novel plants which thanked its origin to crosses with different species, will be named after the main parent species. And horticulturists will speak and write about new hybrids in Primula obconica or new hybrids in the Shirley poppy.

It is very evident to anyone who studies the work of those horticulturists and amateurs, who pride themselves upon the production of novel plants, that very often very valuable F_1 hybrid plants are thrown away because they are not in themselves remarkable, but which should have been preserved as the potential parents of the desired novelties. This one point, it seems to us, is where some knowledge of Genetics is useful for the practicians.

But the further we progress, the more evident does it become that the average horticulturist who is working to produce new species of plants, and to ameliorate existing species, is very much nearer the truth with his ideas about variability and its causes than the average genetician. A very striking instance of this fact can be seen in the attitude of the horticulturists toward de Vries' mutation-theory. When de Vries revived the idea that new species sprang into existence spontaneously, without ascertainable cause, the horticulturists slowly began to see mutations. Or rather, and this is significant, they began to write in terms of mutability and to speak of mutations, they began to use the terminology of de Vries. At the same time they applied this terminology wrongly, they simply substituted the new fashionable term for "novelty" or for "new species", without changing their ideas about the origin of these novelties in the least. Whenever a horticulturist by means of judicious advertising, flattery and auto-suggestion is made to believe himself a great man, he may branch out into science, and begin to give his real opinions based upon facts observed instead of the latest fashionable theories. When Mr. Burbank, a typical horticulturist of the old school in California, was visited by de Vries, the latter was able to make Mr. Burbank talk and write about mutations. But when a little later Mr. Burbank was made into "a famous scientist", he began to write books for himself. Being thoroughly convinced of the fact that the variability in his plants, enabling him to make his selections, in the way in which all horticulturists have been working through the last centuries, was caused, not by sheer good luck and spontaneously, but as the result of appropiate crosses made by Mr. Burbank himself, he nevertheless for a time retained the term "mutation" then in vogue. He simply stated, that by crossing species he produced "mutations". What he meant is simply, that by crossing he obtained plants with new qualities, not merely recombinations of existing ones.

A very unfortunate confusion of terminology has resulted from this coöperation between de Vries and Mr. Burbank, for, curiously enough, several very able Geneticians, including Morgan and Davenport now use the term "mutation" in Bur-. bank's sense, instead of in the way the term was applied by de Vries. In the chapter on Mutation we will try to show that, whereas real mutation, in de Vries' sense of a spontaneous change in the personnel of the genes is too rare and too insignificant to be a real factor in evolution, it nevertheless does occur occasionally. If we could be sure, that there did not exist real cases of spontaneous loss-mutation, it would be perfectly allowable to use the term "mutation" for "individual with a wholly new character" in the way it is used by so many American authors after Burbank. For it is evident that almost all the cases of the production of a new biotype with some wholly novel characteristic, and which are claimed by de Vries as due to spontaneous changes in the genes themselves, are due to recombinations of genes.

The very simplicity of the earliest cases of "germical analysis" has resulted in a temporary strengthening of the position of those authors, who held fast to the Weismannian conception of heredity as a transmission of determinants for definite organs, and definite characters. When it became apparent that in the majority of instances, there is no simple segregation of the off-spring of hybrids into classes with clear-cut characters corresponding to the parental ones, the result was, that a distinction was made by several authors between two kinds of inheritance, alternative inheritance and blending inheritance.

It is apparent, that, if we consider the characters of the organisms we cross, their phenotype, the inheritance of almost every character envisaged is blending. Unless two individuals are very closely related, but differ mainly through presence and absence of one or two genes which have a very pronounced influence on some very marked character, the off-spring of two diverse individuals is generally intermediate. This fact was felt as a very uncomfortable one by the first "Mendelians", who almost to a man conceived of their inherited factors as of direct determinants for the characters studied. The first attempts to bring these results into line with the hypothetical purity of the factors consisted of trying to analyze the differences between the types crossed into a great many separate pure unit-characters. This could be done in a great many instances, but in many other cases the attempt had to fail. A number of cases remained in which, the difference between two individuals was essentially a difference in just one character, but in which nevertheless the inheritance of this character was not alternative.

This difficulty was necessarily felt as a very serious one by all those authors who held on to the determinant-conception of genes. If there should be a direct and reciprocal connection between a determinant for some character and this character, it would be obvious that the production in generations following a cross of individuals with a character intermediate between that same one in their parents would tend to prove that the corresponding determinant for the character could vary in quality, and could exist in different conditions. Of late years we have read a good deal about the difference between so-called qualitative and quantitative characters. It was generally conceded that the inheritance of qualitative characters such as colour, or presence and absence of organs, followed Mendel's law, but that the case of quantitative characters constituted a difficulty.

What, really, does this difference between qualitative and quantitative characters consist of? If we examine the difference, we find that it is only one of degree after all Qualitative differences are such as result when in the material we are working with, one gene happens to have a striking influence on some character, and when this gene is not present in the genotype of all the individuals. In such cases, the distribution of such a gene over the germ-cells of a heterozygote can easily be observed, as its presence in the zygote can always be detected

by a mere inspection of the characters of the individual growing from it. Now such cases, in which a clear-cut segregation
of the off-spring of heterozygotes unsuitable test-matings goes
absolutely parallel with the segregation of a gene must be relatively rare. For it is evident, that such a distinct segregation
into two classes of individuals can only be expected in those
cases where the influence of the gene is relatively very great.
Black and agouti in rodents are alternative, qualitative characters. This means, that in a mixed group of only blacks and
agoutis, the presence or absence of the gene which by its presence or absence produces the difference, can be immediately
detected by a mere inspection of the colour of the animals. In
this case, it is at once clear, that this fortunate circumstance,
which so simplifies germinal analysis, is not so much due to the
nature of this gene, as to the relation between the influence of
this gene and that of the whole further genotype of the members of the group. This same gene, which in this case, in animals of this genotype, differentiates blacks from agoutis will
have very much less influence in other groups, and even none
at all in some. In light silvers, and mice of very light shades
generally, the presence or absence of this same gene has so
little influence upon actual colour, that there is no possibility
of grading the off-spring of heterozygotes into two classes.
Only test-matings with suitable animals can show the presence
or absence of our gene in animals with doubtful colour. The
difference between animals with and without the gene is not as
great as differences in shade due to age, sex, moulting, general
health. Presence or absence of one and the same gene in the
same material may result in some cases in qualitative, in others
in quantitative differences, according to the rest of the
genotype.

A comparison of the influence of one and the same gene in
groups of different constitution shows, that in one case ; res-
ence of one gene in some and absence from other members of
the group may result in a distinctly discontinuous variation, in
another case in a continuous variation.

The only really satisfactory proof of the existence of a gene consists of showing its distribution over one half of the gametes produced by a heterozygote, and its distinctness from others. In the most favourable cases, this distribution is seen directly, as a Mendelian segregation of the off-spring of a hybrid into two classes. In other cases, where the influence of the gene is not as striking, test-matings will have to be made to determine the genotype of the off-spring of heterozygotes.

Another difficulty in the path of germinal analysis is given in the fact, that only very rarely we can so choose our material as to reduce heterozygosis to the gene we want to study. Ordinarily we can never know, how large the number of genes may be for which two individuals we are crossing are not identical. Even in those cases where the heterozygosis of the hybrids is great, the distribution of such genes as have a very striking influence on the development is easily studied. But it is evident that in very many cases several genes for which the parents differ may influence the same character.

When we are dealing with two distinct genes which influence the colour of our rodents, it is often relatively easy to study the distributions and mutual independence of both. But in other cases the effect of the same two genes may become so much alike that the analysis becomes hopelessly difficult. The only way out in such cases is to reduce heterozygosis, to study A in families pure in respect to B and vice versa.

When we are studying the distribution, and the effect of a gene, which in our material has a very strikking influence upon the development, we can afford to neglect for the time-being the influence of other genes, and of environmental factors, so long as these do not influece the character we are interested in, in such a way as to interfere with our analysis. But it is going altogether oo far to expect this great simplicity wherever we want to study a gene, or to conclude, from the fact that such a simple direct relation does not exist in every case between a gene and a pair of contrasted characters, that there is a fundamental difference between the genes themselves.

Size, weight, is a typically blending, quantitative character. It is evident, that almost any gene which has any influence upon the development of an animal or plant at all, must somewhat influence the final size, stature, height of the individual. And for that reason, we must expect that, whenever we cross two individuals of different strains, and when we ignore the number of genes distinguishing their genotype, we must expect those genes which influence size, and for which they are not identical to be numerous. This means that we can only expect to find a very great variability in size due to variable genotype in the second generation of almost every cross. Especially should this be true where two strains crossed differ themselves in size.

In all those experiments in which the size of the individuals of an F_2 population has been studied, derived from hybrids between species of different s ze, great variability in size has been noted. But very often the fact that no clear-cut 3: 1 or 9: 3: 3: 1 segregation has been observed, has been taken as proof for the theory that the genes which "determine" such "quantitative characters" are different from other genes, and are in themselves variable.

We must remember, that several cases have been noted in which two different sizes in one population behaved as alternatives, and where size was inherited as a pair of Mendelian unit-characters. We should find such cases whereever a single gene has a very marked effect upon the development, and so on the final size, when the variable effect of environmental variations and heterozygosis for other genes does not interfere with our observations of this presence and absence of the gene,

We have the case of the dwarf guinea-pigs, studied by Miss Sollas, that of the cupid sweet-pea, that of recessive dwarfs in beans and dominant dwarfs in Shirno wheat.

If our explanation of blending inheritance is correct, if absence of evident segregation is due to complexity and to the relative magnitude of variations in size induced by differences in the evironment, it should be possible to find evidences of Men-

delian segregation of size-influencing genes even in material
where their presence cannot be demonstrated at once.

With this object in view we started an extensive series of
breeding-experiments on size in mice. We obtained very good
material in the Orient, very minute domestic mice which
proved extremely pure in respect to shape and weight. For the
other species we were fortunate to obtain a strain of large white
mice which had been purely bred by Dr. T. B. Robertson of the
University of California for his experiments on growth. These
mice were very pure in respect to weight, and their standard
weight had been tabulated for both sexes and all ages from one
week up to time of natural death.

This made it possible to use these figures as a standard. Instead of expressing the weight of any or our animals in grams,
which would have restricted the use of this weight to comparing it to that of animals of its own age and sex, we expressed
the weight of any given animal in procentages of the weight of
the standard animals of that sex and age. This made it possible
to compare the weights of brothers and sisters directly, and
also that of animals of approximately the same age.

This is not the place to give more than brief outlines of our
problems and our results. Let it suffice to say that our problem
was threefold. In order to show the existunce of one or more
weight-influencing genes, we had to control as far as possible
the influences of the environmental factors, of age and sex, and
of heterozygosis in respect to other weight influencing genes.

By keeping all the animals in the same kinds of cages, in uniform temperature and on the same diet, all through the experiment we tried to eliminate as far as possible the effects of a different environment on weight. Differences due to sex and age
were equalized in the way indicated above. Rests heterozygosis.

The F_1 animals are necessarily heterozygous for all the genes
which are not common property of both parent species. Segregation of a weight-influencing gene over the gametes of these
animals can be expected only if the influence of this gene is rel-

atively enormous. As can be seen from Fig. 13 which illustrates the weight in pro-centages of a group of F_2 animals, there is a very strong indication for the existence of one or two such genes. The variation curve, far from being smooth is decidedly two-topped. As can be seen from Fig. 14 the shape of the variation curve is not the same at different ages.

From this it is apparent that at different ages, different genes play a weight-influencing role.

When hybrids of this first generation are mated back to either parent, we can expect much clearer evidence for factor-segregations. For, whenever we make this cross, we can limit our attention to those genes which the other parent contributed to the list for which the hybrids are heterozygous. And when this mating-back to either parent species is continued for several generations, eventually purity will be reached, the result of the series will be an individual pure in respect to the genotype of one parent species. When in every succeeding generation we pick one individual at random, we know that that individual has one chance in two to be homozygous for every gene in respect to which its hybrid parent was still heterozygous. In other words, heterozygosis is reduced one half in every succeeding generation. Eventually in every series a point will be reached, where the hybrid individual mated back to one of pure strain differs from it in just one gene. Whenever this gene happens to have an appreciable influence upon the development, and therefore upon weight, this heterozygosis may show itself in a Mendelian segregation into two weight-classes. Whether we ever succeed in demonstrating such cases is a matter of pure chance. The only thing we could do was to have as many separate series of back-mating experiments running as we could afford.

Two sets of series only were continued for any length of time, namely those, where we mated back hybrids between Japanese mice and large whites to the large Robertson strain, and where we did the same with descendants of a Chinese mouse. I have selected as an illustration the offspring of two sisters in the

Chinese series, which were born in one litter and had the same breeding. These animals had one F_1 and one Robertson parent. Although the data are not sufficiently worked out to admit of more than a general statement, it is clear that they show evidences of a complex factorial segregation, which is becoming more and more simple the further the genotypic constitution of the hybrid stock is made to conform to that of one of the parent-species by repeated back crosses. (Fig. 12, 13, 14, 15.)

This brings us to the work of the Drosophila specialists.

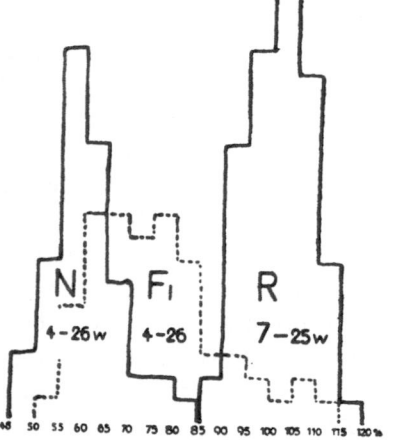

Fig. 12.

Variability in weight of 49 Japanese dwarf mice, 75 large whites, and 60 F1 animals. Weight expressed in percentages of standard weight of white mice as published by Robertson.

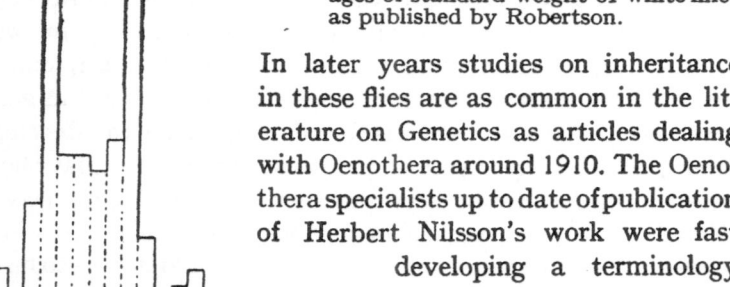

Fig. 13.

Variability in F2 from the cross Japanese dwarf large white mice weight expresses in percentage of standard weight. Males and females at four, five and six weeks.

In later years studies on inheritance in these flies are as common in the literature on Genetics as articles dealing with Oenothera around 1910. The Oenothera specialists up to date of publication of Herbert Nilsson's work were fast developing a terminology and a technique of their own, which tended to cut off these authors from the Geneticians interested in questions of a more general nature. The awe with which we outsiders looked upon the work done by de Vries and the botanists following him, gradually began to turn into something like irritation when we saw these authors plunge deeper and deeper into their specialty. We saw how they accepted the most startling, and to our mind exceptional phenomena, as something to be taken for granted, and how they kept on piling up astonishing facts with a decided unconcern for the possibility of an explanation in terms common to the whole of Genetics. The paper of Nilsson removed a great deal of the tension, as it showed us the possibility of understanding something of the mysterious thing which we were all hoping the specialists would investigate, but which they had until the time only taken for granted.

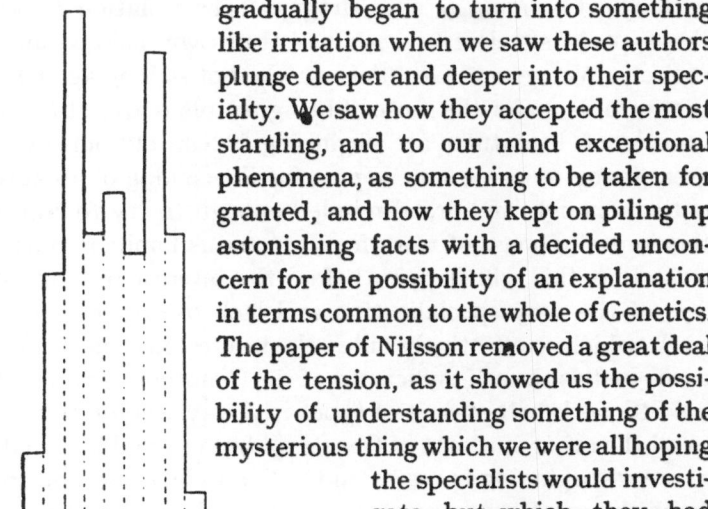

Fig. 14.
Variability in F2 from cross between Japanese dwarf x large white mice. Weight expressed in percentages of standard weight. Males and females at 11, 12, 13 and 14 weeks.

completely as de Vries and the other Oenothera specialists. Already they are writing in terms which are not the terms of the general Genetician, and which evidently are perfectly clear and reasonable to the initiated, but astonishingly unfamiliar to a great many of us.

It seems impossible to speak of what happens in their cult-

There is some danger, that the Drosophila specialists will cut themselves off from the possibility of intercourse with the other Geneticians as

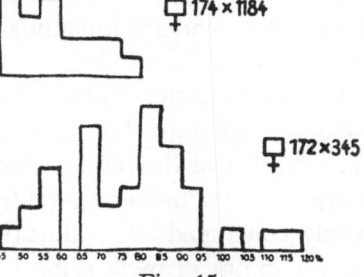

Fig. 15.
Two sisters, 174 and 172, bred from F1 x large white, and their offspring from large white males. Weight expressed in percentages.

ures, in terms of what we are used to think about in our work
with wheat or mice or snap-dragons. The isolation in which
these specialists elect to work is of their own making, but not
entirely. We, general Geneticians, who not so long ago felt uncomfortable under the sense of being unable to digest this new
branch of literature, are beginning to console ourselves. A
great many of us are now hoping for the coming of an author,
patient enough, and broad-minded enough to try to translate
this imposing mass of work for us in terms familiar to us. We
trust that this will happen before the interest in Drosophila
dies of the very specialization and lack of relation to other
work, which so effectively killed the interest in Oenothera for a
time. Just as the self-sufficiency of the Oenothera-authors, who
used to work with names of types in a way wholly unfamiliar
to all the other students of Genetics, shut out and irritated us,
so are we beginning to resent, in the Drosophila-authors, not so
much the great apparent complacency with which they regard
each others diagrams of chromosomes, but the evident lack of
sense of proportion shown in the work. We mean here the lack
of interest in the work of the others, and in the relation of such
work to fundamental questions of common concern. Nobody
objects to specialization, but a great part of our beginning feeling of resentment arises from the fact, that these specialists
evidently do not wish to leave speculations upon the broad issues of Genetics to such men as Bateson, Morgan, Baur and
Johannsen, who are unhampered by specialization in the material chosen.

   One of the terms we find recurring in the litterature on Drosophila is the term "mutation." It is important to emphasize
that this use of the term is wholly in the *Burbank* sense, as synonymous to "novel character" and that it should not be
rashly concluded, that mutation in the sense of spontaneous
change in one or more genes is a rather common phenomenon
in this material.

   As we will try to show in the chapter on mutation, it is next
to impossible to prove the fact of a real mutation having oc-

curred. In autogamous plants it is hard enough, but in animals
and allogamous plants the proof of a real mutation can be given
only under very favourable circumstances, which make all the
necessary test-matings possible. It is quite clear that Morgan
actually distinguishes between Mutation in the sense of de
Vries and in the sense of Burbank and Davenport. But it is
evident that some of the other workers with Drosophila are not
aware of the distinction. From the Weismannian standpoint,
and looking upon genes as determinants for characters, and
upon "unit-characters" as determined by one or a few genes, it
is evident that the spontaneous production of an individual
which is shown to have a clear-cut dominant and new unit-character proves a dominant mutation in de Vries' sense. And
such a mutation would imply either the creation of a new gene
or the change of a "locus" into something different. We have
seen, that Davenport does not hesitate to assume this sort of a
progressive mutation to account for dominant "mutations" in
poultry.

Crossing in the widest sense, mating between individuals not
identical genotypically, will produce those new dominant characters, these new mutations in the sense of Davenport. And if
we find out the facts in cases of the production of a new dominant "unit-character" in Drosophila, we will always find that
such dominant "mutants" happened to originate in cultures
which were not inbred for a long series of generations. The fact
that the dominance and distinctness and novelty of a "mu-
tant" character are looked upon as sufficient proofs for a mutation, is evident from the very tests to which animals with the
new character are submitted, As an example we want to quote
a few sentences from the first lines of a recent paper by Otto
Mohr.

"A female, just hatched, having wings serrated or not-
"ched at the end, was found in the purple stock-bottle,
"October 7 1918. In order to test whether this character
"was due to a new mutation, the female was mated to
"males from an unrelated stock. The majority of her F_1

"daughters has typically "notch" wings, etc. It was possi-
"ble to ascertain that the purple stock had not been con-
"taminated. The character was accordingly considered
"as due to a reappearance of the old notch gene".

As can be seen from this sample, the new character is un-
questionably identified with a gene. The gene itself is called
notched. This is quite typical of the publications on Drosophila.
The same characters appear quite frequently and apparently
spontaneously. Loss-mutations are followed by return-muta-
tions, the "lost character" reappearing. Such cases are often
casually alluded to without much reference to the fact that,
should they actually correspond with loss and recovery of
genes, or with change in dominance of genes, the bearing upon
our theoretical notions about genes should be of the utmost
interest.

So long as a stock is not genotypically homogeneous, that is,
so long as the potential variability of it is not zero, both recessive novelties and dominant novelties can be expected to turn
up in it, and it depends largely upon the care with which the
material is analyzed and studied, how great the number of
discoverable "unit-characters" is.

How is it possible to account for the enormous genetic variability in this Drosophila material, if we do not have recourse
to the mutation-hypothesis for an explanation? Is it possible
to explain the origin of the new characters wholly on the basis
of recombination of Mendelian factors, genes? It is plain that
the difficulties in the way of such an explantion are made to appear unnecessarily great by a strict adherence to the determinant-conception of heredity. We have had many an argument
with American geneticians about the use of mnemotechnic
symbols to denote particular genes. To a great many people it
appears that, when it is demonstrated that a given gene produces the difference between black and white animals, calling it
"Black" or B helps to memorize this action of the gene. And
they are of the opinion, that this system of naming genes after
characters does not produce the confusion which we fear it

does, and does not lead people always to think of the colour black, when they see the symbol *Bl*, or worse, to think of the gene *Bl* whenever they see black colour. The most prevalent argument in favor of the use of terms for genes that recal char-racters, is, that it makes the papers on factorial analysis easier to read and understand than those in which the authors follow Mendel's system of denoting genes by non-committal letters of the alphabet or by numerals. This is a question of taste. Person-ally, we find it exceedingly difficult to dig down to the facts hidden under the laborious terminology of the work of the au-thors using this system, even of those authors like Plate and Morgan, who are obviously not themselves confused. And a great number of papers by recent investigators working with Drosophila are so hopelessly confused in their use of such terms as gene, character, locus, mutation, that it is hardly worth while to try to get a picture of what may have happened in the experiments described.

We cannot help thinking, that an author who consistently writes about a notch gene must eventually come to look upon this gene as upon a determinant for the character, and upon the character as bound up with the gene.

What we try to study in work of this kind, is after all the mechanism of heredity and segregation, the relation of the different genes to each other. It is possible to read a paper on factorial analysis and to concentrate one's attention upon the mutual relations (linkage, coupling) of six genes called A, B, C, D, E, and F, without verifying at the time in what particu-lar way, and in what combinations these genes demonstrate their presence in the material. Speaking for ourselves we are very much interested in the mutual relations of the very many genes demonstrated by Morgan and his pupils in Drosophila and only very little in the particular characters of the flies which vary certain combinations of these genes. We would be overjoyed, if someone would transpose the facts in some of the most interesting cases and would denote the genes sim-ply by numbers, numbering them from one to four-hundred-fifty, because it would give us a chance to study the relations of genes 4, 28, 34, 36 and 329 without being detracted by symbols which stand for fantastic characters that result in certain types in the presence, or even in the absence, of such genes.

The behaviour of the genes in heredity, their relations during segregation are things to be studied by themselves. These are the things which readers of papers on Drosophila are interested in. The other geneticians, reading about this work are, for the time being, content to take the specialists word for the analysis of the counts of his flies.

On the other hand the specialists working with the material have to know, what each gene stands for. From personal experience we know, that bare numbers very rapidly associate themselves with memories of peculiarities of the material. In recalling our work with squashes the number 161 brings to our mind a particular type just as clearly as the name "Miracle" does, and in our pedigrees of our mice numbers 2112 of the old series and 147 and 148 of the new series recall particular animals quite distinctly.

We should clearly distinguish between these two desiderata, making the data easy to work with for the experimentators, and making them intelligible for the other geneticians, who are expected to read them.

If we know a gene which is present in black animals and absent from brown ones, and a second one which is present in black-eyed but not in pink-eyed, we can study the relation of these genes to each other just as easily, if we call them A and B or 15 and 16, as if we called them Bl and Bl^e. If we call them A and B the facts are intelligible to everybody. If we call the first one Bl, this abbreviation suggests "black" to the man who uses the abbreviation. A French author would call this gene N, a Scandinavian or German would prefer to call it S, we Hollanders should call it Z.

But, after all, there is no real blackness anywhere about this gene. It does not regularly produce blackness. In a very definite combination of a dozen or so other genes which we

know, and a few hundreds which we do not know, its presence
or absence makes the difference between a black and a non-
black individual. We can study its inheritance in groups of
animals where its presence does not produce black colour at
all. In all those groups, in all those genetic formulae of animals with and without this gene, there is no real advantage to
be made to remember that this is the gene, which would produce black colour if genes number 2 and 3 and 4, 5, 8, 9 and
12 were present and numbers 6, 7, 10 and 11 were absent. Calling it 1, or 13 would be simpler.

Calling a gene by a name taken from some character which
it may help to produce in certain combinations, has its grave
dangers. It produces an association between the idea of the
character and the gene and it makes for a false simplicity in
the work. Some people are even to-day wont to think of genes
as of determinants for particular characters. To take a few
examples from the work of the Drosophila authors Notch and
Purple and Peach are obviously names of characters of certain flies. But the names are also used to denote certain genes
which are thought to be always present, or always absent
when these characters are seen. So long as N and P do not
pretend to be more than our A and B, symbols for genes demonstrated in the analysis, the use of the two first letters
rather than the two others is wholly a matter of taste. But
when we read of the localization in the chromosomes not of
genes A and B, or genes N and P, but of Notch, and of Peach,
we begin to see that the use of these names for genes has hampered the author in his thinking about these genes and their
re.ation to the characters of the flies, which he also calls Notch
and Peach.

From experience we know how very laborious is the work of
disentangling the several genes which can be demonstrated
in our material.

When we know six mutually distinct genes, the amount of
work to be done, before we are satisfied that a seventh is really
distinct from all the other six, is apalling. And yet we cannot

be said to have proced the existence of a seventh gene, unless
we can prove by all the necessary breeding tests that this
number seven is not perhaps really number three, or that what
appears as a distinct gene G is not really a combination of B
and E.

We firmly believe, that, if we want to do such work, we
must never lose sight of the fact that the action of a certain
gene upon the development is not always the same, but differs
with the set of other genes present and with the action of the
environment. If we meet a gene for the first time in yellow
mice, and we find that it distinguishes these yellow mice from
non-yellows, it does not follow that all yellow mice are yellow
because of the presence of this gene; or that all mice with this
gene must be yellow. In fact, we positively know, that mice
without this gene can be yellow as a result of different combinations of genes in the absence of the gene studied, and also,
that mice with the gene, may be albino or sable or very faint
pearl grey. It is bad enough to use the letter Y for this gene
(or G, J, K, according to nationality), but if we find an author
writing about the localization of Yellow in the chromosome,
we feel convinced that he should not be trusted very far in an
analysis of the other genes in the same material. For it is evident that, if such an author discovers a gene influencing the
number of scales on the tail, or the curvature of the claws, it
will not occur to him to make the necessary tests to prove that
this new gene is not in reality his gene Y.

Every new gene can be accepted as such only, if its distinctness from every other hitherto described gene is sufficiently
demonstrated. It is clear that, according to this standard, only
a very slight proportion of the genes name in Drosophila can
be accepted, and that the proofs which most of the authors
require for the acceptance of the novelty of a gene are absolutely inadequate. If we did not know from the work with rodents and with peas, how very striking can be the new characters produced by novel combinations of genes, and how very
often it is wholly impossible to judge of the particular com-

bination of genes which produced the new colour, we could easily be led to accept a novel character which showed a monofactorial segregation in crosses with "wild" stock, as proof of the existence of a hitherto undescribed gene.

So long as on one hand the "mutations" in Drosophila are accepted with so little reserve, whereas on the other hand the cases, in which new characters, new colours, new shapes have been shown to result from novel combinations of known genes, have been so very little worked out, and only in the crudest possible way, we are justified, we think, in attributing a great deal of the startling departures of this material from other material, and of the constant need for more and for more complicated subsidiary hypothesis to this peculiar terminology, to this system of naming genes after novel characters.

For this system evidently promotes in the younger authors a tendency to look upon the relations between genes and characters as direct and reciprocal, and to andre-estimate the necessity of rigid proofs in the work of factorial analysis.

It seems probable that a certain amount of the genetic variability in Drosophila studied in recent years is due to irregularities in the usual constitution of the chromosomes, and therefore to real mutations. Shull has very clearly shown how "crossing over" between chomosomes, including "longitudinal crossing over" could explain simultaneous duplication of genes and loss-mutation, without recourse to real spontaneous loss or the less easily conceived spontaneous acquisition of genes. Disturbances in the usual behaviour of chromosomes have been made very probable in Drosophila. On the other hand it is clear, that the evidence for the existence of so many distinct genes in Drosophila, distinguishing animals with different characters, is wholly inadequate, and that the possibility is not excluded, that a great part of the unexpected behaviour of characters in crosses, which is now met by a novel hypothesis, depends upon the circumstance, that new characters are not expected to originate by novel combinations of partially known genes.

And this is especially true of the new dominant characters. New dominant characters, resulting from a cross new in the sense of not being observed before, and not being expected as a simple combination of two known characters, should be very common in such diversified material as these flies. We find no data in the litera ure on cases of this kind. And the absence of such data, observed in connection with the frequency with which dominant "mutations" are described, makes it more than probable that most, if not all these "mutations" are due to novel combinations of genes already studied, and included in the material. The method of firmly associating particular genes with particular characters must almost inevitably bring about this result. If an author has studied the mutual inde-pendance of genes A, B and C, he may not feel sure that a novel dominant character suddenly cropping up is due to a new gene D, before he knows that this new character is not a very common result in individuals carrying A and B, or A and C, but if one works with genes named after definite char-acters, he may think it absurd even to conceive. The possibil-ity, that a novel character fitly called "pushpin," may result from a genotype including genes "duckfoot", "gold" and "sticky" but lacking "chestnut," and instead of making all the possible combinations of all his genes studied so far, he will contend himself by christening a new gene "pushpin" after the novel character of the fly found in the "duckfoot" stockbottle.

So long as the possibility exists of naming one and the same gene six times after characters it helps to produce in different combinations of other genes, so long is it unneccessary to re-gard the production of striking dominant novelties in this material as evidence for progressive mutation in the sense of de Vries.

It must be understood of course that the tendency among some of the American authors is to use the term "mutation" as synonymous to "novelty", in the sense of Burbank, rather than in that of de Vries. In those cases however, it should be made perfectly clear, that the term is used in this way.

REDUCTION OF VARIABILITY.

FROM a Biomechanical standpoint heredity is the transmission of genes, which under certain circumstances can influence developmental processes and in this way final qualities of the organisms. This conception of heredity is fundamentally different from the de Vries—Weismann conception of the process as a transmission of pangens, determinants, which would each directly call into being, directly determine a corresponding organ or quality. This last conception brought with it the necessity of assuming, that such determinants could exist in a definite latent, dormant state, namely in all those instances where we knew a certain inherited thing to be present without the corresponding quality with which it was commonly found associated, showing itself.

If we look upon genes simply as upon substances, which by their presence act upon the course of definite developmental processes, growth-processes, we need not assume that they are dormant or latent in those instances in which the process they can influence does not take place.

If we take this view of inheritance, we can understand how in a species of plants or animals large numbers of genes may be common property of all the cells, which genes in organisms of this particular biotype do not actively participate in the development.

In rats we know a gene, which, when present in coloured animals, makes otherwise black animals agouti. In albino rats, this same gene, though it has no influence upon the colour, is nevertheless transmitted in the same way as in families of animals in whose development it plays an active rôle. It is regularly distributed over one half of the number of germ-cells produced by individuals impure for it, as can be readily seen by subjecting all the members of a family of albino rats to a suitable test, namely to a cross with pigmented animals lacking the gene in question, such as blacks, or pink-eyed creams. A gene may influence two different developmental processes in the life of the same individual, and if one of the two processes is so changed that it does not any more come under the influence of our gene, the fact that the gene itself is not latent or dormant, is seen from the fact that its action on the other process remains unimpaired.

If we should ever be called upon to give as complete a list as possible of the genes for which a given family of unpigmented rats is impure, we should have to make suitable test-matings, to decide whether or not the individuals were pure for such genes as did not happen to show their presence.

We propose throughout this book to use the term *Total Potential Variability* for the number of the genes in respect to which an individual or any group of individuals is not pure, homozygous. In doing this we place ourselves upon the standpoint, that in inheritable variation we are concerned with the influence of the genes exclusively, and that the individual genes are qualitatively stable (Law of Johannsen) so that variability, and potential variability becomes synonymous with genotypic impurity.

In practice, it is obviously impossible to determine the exact number of genes for which an individual is not pure, or for which not all the individuals of a group are pure, and therefore to put down the exact number, expressing the total potential variability. But even so, this number is a definite one, even if we do not know its magnitude, something we can work with in certain ways. We can investigate which processes make the number larger, and which things reduce it, and in what measure they do reduce it.

In our investigation of the methods of evolution, we see at once that we are everywhere concerned with two mutually opposed tendencies, principles. Darwin called these variation and heredity.

On one hand we see causes which make the total potential variability of the groups larger, and on the other hand we find things which tend to reduce it, cut it in half or even eliminate it.

The potential variability of a group of organisms becomes larger if into this group individuals are taken up, which either possess a gene or genes not heretofore present in any member of the group, or which on the other hand lack genes, common property of all the members of the group. Here of course we see the difference between variability and potential variability, because the last is concerned with all genes, be they factors in the development or not. Two animals or groups of animals may have the same characters, and the same variability, and yet they may differ in possession or lack of one or more genes, which in these types have no effect. In such a case their hybrid off-spring may not be more variable, and at least in one generation variability may not be increased. The potential variability however, did increase, and this may come to light in the second generation, or even much later in the behaviour of some descendants off-spring from yet another cross.

We saw that mutation, which means loss-mutation, can at the most play only a very insignificant rôle either in the heightening of the potential variability, or in its decrease. With this insignificant exception then, we can focus all our attention upon crossing in the widest sense as the cause of variability. And in this chapter we are concerned with the second part of the evolution question, what causes reduced variability?

This was the question with which Darwin was mainly concerned. Given a certain variability, how do species finally lose this variability and become pure? We know that the answer of Darwin was, that selection causes this reduction of variability. As there will in every group be individuals which are better fitted than others to survive, and as it will be these individuals which on the whole survive, the whole group will gradually tend to change in this selected direction.

In so far as variation brings about a greater or lesser usefulness of some organ, some part, there is not much to be said against this reasoning. But it rather implies, that any quality for which we see that a species is now pure, must have, or at least must have had, its use, or must have been correlated with something useful. And this consideration has led a great number of authors always to look for the usefulness of every trifle, they could discover about plants or animals.

If a spot on the wing of a butterfly makes it a little less conspicuous, it is the intention of this spot to bring about this inconspicuousness. If, however, the spot makes the animal rather more visible, then the spot is there to make the animal conspicuous. If we observe the stripes of a zebra, we think these stripes make the animal invisible. If next we observe that our zebra cannot keep its switching tail from moving continually, and giving him away, we decide that this nervous trick is an adaptation to a fly-infested country. A bird looks like another bird to which it is not related, and forthwith we declare that the one must imitate the other. Dewar and Finn give a whole list of instances in which two birds resemble each other very much more closely than the classical examples of imitation, but in which the two members of such a pair inhabit different continents and never had any relation to each other. Almost any bit of coloration of any organism can be said to be useful in some way, if the good faith of the naturalist is only sustained by a competent imagination.

William Ritter has repeatedly pointed out the sterility of this hunt for the meaning of everything. If we see that the occiput of a bird is black, why, it is black, because it is not blue or pink, if it were not black it would have some other colour, colour it must have.

It is very apparent, that if two birds can coëxist in one environment and be so nearly alike in habits and nesting-place as European the blue tit and the great tit, we cannot invoke the usefulness of the sky-blue markings of the smaller bird as the cause of the fact, that all the individuals are pure for the combination of genes which results in this colour, any more than we can imagine how the black markings of the other tit are useful to it in any way. How can the striking black and pearl grey pattern of the hooded-crow be useful to the animal, if we know that it only differs in this colour from the black crow, whose territory touches his, which has the same habits and breeding-places, and inter-breeds with it where they coëxist? Of what use is the black tail-tip to the stoat, if the weasel can thrive without it? Why is the tree-martin pure for its yellow throat, and the house-martin for its white one?

When we see how the purity, the stability of characters in species is as great in respect to trivial things as in respect to important characteristics, we are made to think. The usefulness of a small thing may escape us, and on the other hand we may over-estimate the usefulness of a seemingly important thing. But even so, it is obvious, that there are untold characters for which species are pure, and which cannot possibly be accounted for as useful. In such cases natural selection cannot be depended upon to furnish the clue. Conscious selection in plant-breeding work is a process which on the whole is far more severe than natural selection. It is remarkable, that in new plants or animals, which have been subjected to a rigid selection in respect to one or several useful characteristics, purity has not only been attained for these characters, but equally as well for the most trivial things to which the selectionist never even gave a thought. We are speaking of the shape and the arrangement of small hairs on the seed of some cereals, of the colour of the eggs in fowls, of the juvenile colour of rabbits, the shape of the leaf in coffee, the colour of the leaf of sugar-beets.

We cannot fail to understand that there must be something besides usefulness to cause stability, purity. And if we find a way to account for specific stability, which does not in every instance take into account fine shades of usefulness, in other words, if we find the nature of the process which causes purity during the process of natural or artificial selection, we shall have made an important step toward an insight in specific stability.

It is rather obvious that it is not selection itself which causes purity, but some process which accompanies selection. Therefore, if we discover what there is in selection to cause purity, decrease of variability, we can pass on to the question whether this process acts independently of selection, and therefore independently of usefulness.

It is here that the proposed term Total Potential Variability comes in useful. We have used this term to express the number of genes for which a group of organisms, or one single individual is not pure.

The simplest instance of a group of organisms exhibiting a small potential variability is the case of two or three plants which all have the same set of genes, which we will together call X, with the exception that each has an additional one, which is lacking from the others. The constitution of the three will therefore be X—A, X—B and X— C. To simplify matters, let us suppose that there is no crossing; we are dealing with self-fertilized plants, or even with asexually propagated plants. The total potential variability in our group of three plants is three. If we multiply these three plants, the potential variability remains three, that is, it remains three just as long as in every generation at least one plant of each type is reproduced. And if for any reason any one of the three types, X—A, X—B orX—C dies out, or is not included in a new group, a new colony, that new group or new generation will have a potential variability which has dropped from three to two. Any new colony which originates from our mixture, out of a few seeds or a few tubers, will have the original potential variability if all the three types are included. But if they are not, it is lowered to two, or one, or zero. Any little group starting from seeds or tubers of one individual will in our example be without any potential variability, it will be in every respect pure. And right here the real nature of selection as a factor in purity comes to light. If from a mixture of three pure lines we select one plant with desirable characters, the resulting group will be wholly without genetic variability. But it is clear, that it is not the selection which is responsible for the purity, but only the colonisation from one plant chosen by the selection. Any other cause, which isolates the progeny of one plant from the mixture will have the same effect, will cause the same purity. Hallett and de Vilmorin introduced into plant-breeding the principle of construing in every generation from one single individual, where the plants were self-fertilized. In these plants, as well as in asexually propagated ones, the effect of choosing one single individual is a complete elimination of all genetic variability. Once the plant with the most suitable genotype is found, the further work consists merely of propagating it, and keeping its progeny free from admixture. The production of new wheats or new sugar-canes, by this process of choosing one single individual, illustrates better than any other process, the fact, that it is the isolation which here produces purity and not the selection. It has been demonstrated repeatedly, that any individual in such plants will have a uniform descendance, and on the other hand it is easy to show that mass-selection in the same material will be comparatively ineffective to produce a high quality, and a great uniformity. In comparing mass-selection and individual selection in the work of the plant-breeders we commonly compare two processes, which are more unlike than is implied in the terms. The system of selecting one individual plant, either as such, or after examination of its progeny (Vilmorin's method) is generally followed with self-fertilized plants, with rice, wheat, oats, tomato, pea, whereas mass-selection is the most common method of breeding allogamous plants. This must be mainly due to conservatism. Wheat is still bred essentially in the identical way in which de Vilmorin started breeding it, and sugar-beets are still bred in the majority of instances in the way in which he originated them. Only comparatively recently the plant-breeders have made the discovery, that even in habitually cross-fertilized plants the variability in the off-spring of one plant is very much smaller than that of the mixed off-spring of a greater number of individuals. And so individual selection has under various names been introduced as a working method in breeding-work with these plants. In corn, for instance, the "centgener" or "ear to row" method of selection has become quite popular, and in tobacco a similar system is now generally followed. Even in sugar-beets, in which self-sterility makes even enforced self-fertilization impossible, and where a special, modified system of mass-selection has been consistently followed, we have begun to realize the hand- icap under which we are working. In sugar-beets all the mod- ern seed-growers now use a method of comparing the progeny of a great number of individual plants. We begin to under- stand how the very great variability, and the "degeneration" of sugar-beet-seed is to a great extent avoidable. The total po- tential variability of one individual beet, and ot the descend- ants of this beet is only a fraction of the variability of the whole group. And if we select six families, the progeny of six individual beets, out of all the lots compared, we know, that the genetic variability of these six groups is not identical in all. By mixing the seed, and growing the mixture without select- ion for two generations in order to obtain the required quantity of commercial seed, we allow these six groups to inter-breed. The variability of the resulting seed and therefore of the resul- ting crop is several times greater, than it need have been. By limiting our choice to one single group, the progeny of one single plant, and rejecting second choices, we reduce the po- tential variability, and thereofre the "degeneration" to a fraction of the usual variability and degeneration. Here, again, it is not the selection itself which makes for purity, we select the individual with the genotype which most nearly ap- proaches our ideal, and therefore the characterso f the average individual of the selected group will be as we want them to be. But purity for those characters is caused by isolation.

If we divide any population of plants or animals into two parts, even if these are equally great, it is nearly certain that the potential variability of these two groups will be different. Genes which are carried only by very few individuals will be all included into one group, and if in the original group only one or two individuals are impure for a certain gene, these may be included in only one of the two halves

The potential variability of a group of organisms remains at the same level only for as long as for every gene for which the group is impure there will be individuals, or at least one individual carrying it, and as long as there will be included at least one individual lacking it, or heterozygous for it. For this reason it is only in peculiar circumstances that the potential variability of a group retains its magnitude.

Even in those cases where colonization is random sampling, the sample will seldom be wholly representative. Therefore the greatest chance of finding cases, where the potential variability of a group does not decrease is in well-established, common species. In these species the number of individuals which we find to belong to each remains essentially the same from year to year. Let us suppose that we could make a census of the house-flies within the town limits of a given city on midsummer day. We would very probably find that from year to year the number did not vary materially. Or if we could make a complete census of the pigs kept on a certain date in a given county, we would probably find their number to be essentially the same for that same date on ten successive years. The environment, the economic conditions are such, that they determine a certain number of flies, a certain number of dandelions, of pigs, of rats. Is this a set of cases in which the potential variability remains the same? Or will it diminish? We have to examine the facts a little more closely. Let us choose the case of the pigs. Last year there were a thousand pigs, and this year there are one thousand, and it is very probable that there will be one thousand next year. We bring these figures to a statistician and he calculates from them, that on the average every pig has produced one baby pig. Next we turn to the farmers and ask them whether in reality every pig has produced one young pig in the year studied. We are surprised to find, that quite a number of those thousand pigs of last year died without off-spring. In fact, we find that of the five-hundred males of last census, only ten had any off-spring whatever. So these ten males averaged one hundred children each. And of the five-hundred females, only fifty raised off-spring, they averaged twenty young each. So that, after all, the number of pigs out of the thousand of last year, which produced the thousand of this year was sixty, or six per cent. Last year there where six red pigs among the thousand. Two red young pigs were born this spring, but they happened to be among the herd of a farmer who sold out to the butcher. The chance of a reduction of the variability among the thousand pigs of last generation, the chance that the potential variability of this year's pig-population is smaller than last year's, looks greater when we examine the facts, than when we accept the statistician's calculation.

While we are writing this, our cats are constanly coming and going, bringing in field-mice, Microtis, for the kittens. They live on these mice almost exclusively. The field-mice scamper from under our feet in all the pastures, and hawks and weasels and coyotes do as our cats do. But we know, that when the drought has set in, the breeding will slow-up, and the cats and the owls will begin to make appreciable inroads upon the number of mice. Very soon the cats will begin to take a renewed interest in the kitchen. The statistician will tell you, that the millions of Microtus in Strawberry canyon are descended from as many millions that were living here last June, and that on the average each one of these mice of last year has produced one of the mice of to-day. If we go out to hunt Microtus in December, and find that they are decidely rare animals, we get a different picture. How many dozen actually survive the winter and are the parents and grand-parents of the June millions? How great a percentage is this number to the summer number? Even if the numbers remain the same from year to year, the group of survivors is so small, compared to the group of mice from which chance has selected them, that the potential variability of the group, if there was any, must have been greatly reduced. (Fig. 16).

As to plants, one of us has observed the distribution of Lychnis diurna very closely for a number of years in Santpoort, Holland, looking for hybrids with Lychnis vespertina.

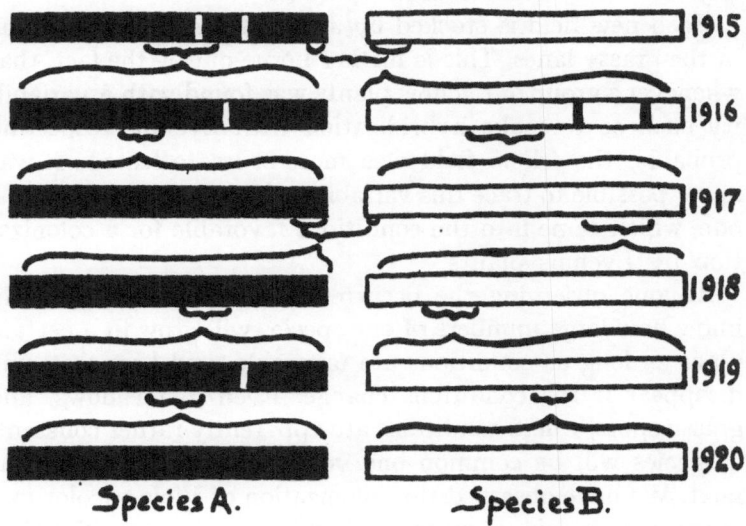

Fig. 16.
Two related species are represented by series of cross sections through the mass of individuals at a number of succeeding generations.
The effect of occasional cross-breeding is shown in the diagram.

Inside the dunes there is an extensive belt of oakwoods. In these woods, the growth which the oaks have made, is cut off at intervals of about sixteen years. These woods consist of lots of about an acre, in different stages of development. When the oaks are cut for fence-posts and faggots, the soil is dug and one crop of potatoes is generally taken. In this year, numerous annual and biennial weeds invade the field. In the second, and third year, these fields are covered with flowers, among which, evening-primroses and Lychnis of both species, take a prominent place. In the third year, the oaks begin to crowd out the weeds, Lychnis diurna disappears first, and Lychnis vespertina a year or so later. There is no continuity. It may well be that in the woods of a particular landowner the

8

number of Lychnis plants remains the same from year to year, but it is certain that for every thousand plants of Lychnis, which we find flowering in one field, there have not been anything near one thousand parents.

Each new field is stocked by a few plants which linger on in the grassy lanes. This is further borne-out by the fact, that whenever a group of Lychnis plants was found with a variability such as to make hybridization with Lychnis vespertina probable, the whole field was more-or-less affected. It was never possible to trace this variability from one field to a new one, which came into the conditions favorable for a colonization by Lychnis plants.

Anyone observing the occurrence of common weeds will know how large numbers of one species will grow in a certain place as long as conditions are favorable, and how they will disappear if the conditions change. Even in meadows, and grassy spots, where conditions are apparently rather constant, a species will be common one year, and rare, or extinct, the next. We have observed the colonization of Viola tricolor in a sandy meadow in Bussum, Holland. In two years the field was blue with the flowers, but after five years the species had completely disappeared. The examples brought forward may seem extreme, but we think they serve the purpose of illustrating the fact that, though statistically a given number of plants or animals descend from a group of the same magnitude of one generation back, in reality they do not. A group of animals or plants descends from a fraction of the number of the individuals of the preceding generation. If we add together the number of individuals of families which have no or little genetical relationship, we may find that on the average one hundred individuals have one hundred descendants, but if we ask the question whether this statistical truth corresponds with a biological one, the answer must be negative.

In such cases as that of the flies, and of the field-mice, the conditions favouring the reproduction of a species may alternate regularly with conditions unfavorable to it.

Every time a small group of animals or plants start a colony, the resulting group will have a limited potential variability. Such animals as have colonizing habits will perpetually start new, isolated colonies from small beginnings. Such animals will be found in groups of which every one is relatively pure. (Lloyd).

The case of the house-rat is a good example. It is not given to much wandering, like the Norway rat. Each farmhouse in France has its own Mus rattus population, and each of these populations is relatively stable. We found the same state of things in Holland, and in Java, as Lloyd observed in British India. In one house all the rats will be grey-bellied agouti, in another house a small colony of white-bellied rats is found, or all black rats.

We saw that in a common, well-established species, where the number of individuals remains constant from year to year, the fact, that the number of plants or animals having off-spring is only a fraction of the number produced, will result in an automatic reduction of the potential variability, an automatic purification of the type.

Let us therefore imagine what would happen in a case, where in reality every individual had only one descendant. What becomes of the total potential variability if in every generation every plant has one daugter-plant? Let us take the case of self-fertilized plants, to avoid complications.

A plant will produce as many gametes with, as gametes without a gene for which it is heterozygous, and in respect to every gene for which a plant is impure, it will produce three kinds of daughters 25 % pure for it, having it, 25 % pure, lacking it, and 50 % heterozygous. In other words, half the number of the daughters of a plant heterozygous for a gene will be impure in respect to this gene, and the other half will be pure in respect to it. As half the number of daughters will be pure for every gene, the chance of one daughter to be pure is equal to its chance to be impure. If a plant is impure, heterozygous for a number of genes, and it produces one daughter, this daughter will be pure in respect to one half of the number of these genes, and impure for the rest of them. A plant, heterozygous for sixteen genes, will have a daughter impure for only eight. The potential variability of an individual plant, which is the number of genes for which it is heterozygous, is one half of that of its mother, and it is double that of its daughter. In every subsequent generation the potential variability is on the average cut in half. The reduction in potential variability may fluctuate somewhat from generation to generation, it may be 40 % for one generation and 60 % for the next, but it will always average 50 %. If we start with a plant heterozygous for sixteen genes, with a potential variability sixteen, the T. P. V. of its daughter, grand-daughter and so on, will be 8, 4, 2, one, and zero. In a few generations the descendants of a plant which is propagated in this way will be without potential variability. In other words, it will be a pure line.

In a population of self-fertilized plants, a mixture of plants of different biotype, each species included will speedily lose its potential variability. In very few generations we will have a mixture of pure lines, every composing species will have lost its potential variability absolutely.

But even if all the composing species are genetically wholly pure, the whole, the mixture is still variable, comprising different strains, different pure lines. Will this variability persist, or will it also tend to diminish automatically?

We have a collection of several hundred species of wheats and other small cereals. If we want to keep our collection intact, we have to take the trouble of separately harvesting a plant or at least a few seeds of every number. Suppose that for any reason we lose interest in the names and origin of my species, and we simply want to conserve them all to use for possible further breeding-work. Can we harvest all our little plots with one stroke of a harvesting machine, mix the seed thoroughly, and sow a sample? Will we keep the collection, or will numbers get lost? It is very obvious, that such a procedure would lose us several of the numbers. If we sow all the seed that is harvested and continue sowing everything, the potential variability of our collection will be kept at its original level. Very soon, however, some numbers would take the upper hand, their seeds would far-out-number the seeds of others. It cannot be expected that a wheat like "de l'extrème sud Algérien" would produce more than five percent of the number of seeds of a "Squarehead." And as soon as sowing all the seed becomes a practical impossiblity, even dividing the bulk fairly in half would probably mean the loss of a few numbers. Let us examine a hypothetical example of great simplicity, and see what will happen if the number of individuals remains the same from generation to generation, in a mixture of species. In continuing our collection, we need only produce one individual of every kind in every generation. If a collection consists of three species, and the number of individuals in every generation is three, the chances that without any selection three seeds will again reproduce the three kinds is $3 \times (3 - 1) : 3^3$ or $6 : 27$. The greater the number of types, corresponding to the number of individuals produced in every generation, the smaller becomes the chance of preserving the collection intact. The greater becomes the chance of losing one.

The chance that a lot of n individuals taken at random will represent n species, or clones, which existed in a mixture containing equal quantities of each is $n (n - 1) (n - 2) (n - 3)$ etc. $: N^n$.

If a population consists of ten pure lines, or ten pure clones, and we know that they are each represented by the same number of individuals, the chance of getting the original diversity if we plant only ten seeds, ten tubers, is $3.628.800 : 100.000.000.000$, or $1 : 27527$. As soon as in such a case one type is not included, it is lost for ever, and not only that, but the relative preponderance of some types brought about by chance alone, will make the chance for others to get lost very much greater than it originally was.

There is a strong tendency to automatic reduction of the

total potential variability in mixed populations, and we saw that there also is a very pronounced tendency in every component species toward reduction and speedy loss of the potential variability, wholly irrespective of selection. Selection will hasten the process somewhat, but we have seen, how it is the limitation of the number of individuals pro-creating, which is the active principle in selection. Selection is only one kind of the several kinds of isolation.

The next case, which we have to examine, is that of the allogamous organisms. We want to know, whether in such organisms there is anything comparable to the automatic reduction of variability we met in self-fertilized and asexually propagated plants and animals.

However much crossing there is in an animal population, or in a group of beets or other allogamous plants, we know that for every gene respect to which the whole group is not pure, not every individual will be heterozygous in every generation. Heterozygous animals will produce homozygotes. And as soon as there are homozygotes (with and without) and heterozygotes for some gene in a group, the potential variability is bound to diminish. Let us take a few examples.

Let us first examine a simple case of inbreeding, a succession of brother to sister matings, in which the number of couples in every generation is one. If the original couple are impure, heterozygous for some gene, A, they will produce three kinds of off-spring, AA's, Aa's and aa's, in proportions of $1 : 2 : 1$. And we can see how great the chances are, that a couple of the children will be homozygous for the gene for which the parents were still heterozygous. There is one chance in two that one of the young pair is homozygous, and one in four that both will be homozygous, but as one may be pure for the presence and the other for the absence of the gene, there is one chance in eight that the new couple will be alike and pure in respect to a gene for which both parents were impure.

However, from this generation on there will be other possi bilities; the new couple may consist of two heterozygotes, or of one homozygote and one heterozygote, or even of two homozygotes, one with and one without the factor. If we calculate these chances, we have to take into account the chance each pair has of producing in its turn a pair of like homozygotes. And in this way, we find that taking together all the possible cases of chance matings of children from heterozygotes, the chance in every generation of one couple having a potential variability of one gene to produce one pair of children pure in respect to this gene is as nine to thirty-two. In other words, if we have a pair of animals or plants with a potential variability 32, the potential variability of one pair of children will on the average only be 32 minus 9, that is 23.

In such a strictly inbred series, it is evident that, once a pair is pure in respect to a gene, whether this means that both have it, or both lack it, this gene has to be reckoned out, in considering the potential variability of the series.

It is, we think, unnecessary to go into details to show how selection hastens the process of purification, the main point is to show that purity is automatically attained, even in the absence of any selection.

The last case which we have to consider, is that of amphimixis, the case of the potential variability in a group of freely crossing organisms. Does the potential variability diminish in such a group? Here we can very easily show mathematically that, supposing every mating to result in an equally large number of off spring, and assuming an uninterrupted steady increase in numbers, the potential variability remains undiminished. Jennings and Pearl have worked out these calculations very thoroughly. However, how does it work out in reality? Is it possible that there is an automatic reduction of variability in freely crossing populations, of the same nature as the very great reduction in self-fertilized organisms and in inbred series, slower, but still appreciable?

In the first place it is obvious, that the multiplication of any group of organisms cannot proceed indefinitely. A few heterozygotes in a mixture of organisms, of which the majority are homozygous will continue to exist so long as multiplication is unchecked. But, wherever the group is continued from a fraction of the number of individuals, or where a colony is started by a few individuals, the chance of the heterozygotes to be included in the group, or to have heterozygous children included, is proportionate to their frequency. Heterozygotes will produce homozygotes, but not the reverse.

The group of organisms chosen by fate to become the parents of the next generation is usually, but always occasionally, considerably smaller than the number of individuals of their species. Every case in which rare individuals, having genes, not present in the majority, or in which rare individuals being impure for, or lacking in genes, common proporty of the majority, happen to be excluded from the number of pro-creating individuals, the total potential variability is lowered.

This, in our opinion, is the most important gain in knowledge which we owe to Mendel's work, and to the biomechanical interpretation of his work. Reduction of potential variability, in other words purity of species is automatic, and not dependant upon any sort of selection. Darwin lacked the necessary key at the time when he needed it most, and when he came into touch with Wagner's work, it could not shake his faith in selection as the cause of stability of species. All the recent work in Genetics, Mendel's law, the things we have since learned about the nature of the genes, the selection experiments with the most diverse material, have shown us that Wagner in opposing Darwin in this fundamental point had the right wholly on his side.

From the way Darwin reacted upon the work of Naudin and Wagner, and from the slight impression Mendel's work made upon Darwin's greatest pupil, Weismann, we are able to see why it did not seeme of great importance to Charles Darwin. But it would appear to me that Wagner would have greatly appreciated it, and could have been trusted to incorporate it into a really logical evolution-theory.

In every single instance in which the proportion of individuals which produce off-spring to all the individuals is estimated, it is found to be small. Moreover, the number of individuals of a given group nearly always vaɩies within wide limits. In extreme cases, such as that of parasistes, it is the rule that one or a few individuals found a colony which may attain to enormous numbers, before it dies out. It is easy to see, how such organisms like malaria plasmodia must form genetically pure groups in very short time. When conditions are favourable, any group of organisms expands in numbers. The proportion of individuals having off-spring is large. When a group gets into adverse conditions, it may die out altogether and leave only daughter-colonies, or a few individuals may survive. In the latter case, the potential variability of the new population must be a fraction of the old one, no matter if the organism is self-fertilizing, asexually propagating, or allogamous.

In nearly every wild animal, which is for some reason or other subject to careful observation, it is found that the numbers in which it exists vary considerably from generation to generation. In not a few instances a family of rapidly increasing animals, e.g. moths, periodically fluctuates in numbers, because of the influence of a parasite. In a few years the moths will multiply amazingly, until the parasites have caught up with them, so that in the next generation hardly any moths are left, and the parasites almost die out because of the dearth of hosts. The moths left are relatively free to multiply for a few generations, until the parasistes catch up with them again, and so on. In both animals the periodical catastrophe leaves only a very small number of individuals, a very small fraction of the number in the preceding generation. After each catastrophe the total potential variability of both animals must be considerably reduced.

If crossing be excluded, any group of animals or plants gradually becomes pure for its genotype, and consequently for its phenotype, its characters, even without any selection. We have tried to show in the chapter on Variation, how the variation we see in groups of organisms is due almost exclusively to geno-variation, and how it is possible for a trained observer to pick out groups without potential variability. Such groups are really remarkably stable and uniform. A trained nurseryman will be able to name correctly hundreds of Appleclones, no matter where he meets them, and that even in winter, by the looks of the twigs. All hyacinth bulbs may look alike to a casual observer, with the exception that some are a little narrower, others a little browner, small differences, which may be thought due to differences of age or soil. But a trained bulb-grower will give you the name of dozens of hyacinths, if you hand him the dry bulbs. And the wheat expert will know his wheats back, no matter where you grew them.

Any species becomes pure for its type, inevidently and automatically, and the only thing which counteracts it, is crossing. Given a certain geno-variation due to crossing, and resulting in the variation of some important organ, selection may come into play and decide the ultimate genotype for which the group will become stable.

If we find that practically all the house-rats in Holland are black, we may think that it is advantageous to them to be black, and if the house-rats in Java are mainly agouti, we may think that to be agouti is an advantage for rats in Java. But if we see that in the same city, small colonies of house-rats are nearly always homogeneous in respect to colour, and different one from the other, we must see that it cannot be the usefulness of any colour which determines the purity of the group in respect to it. And this is what Lloyd found in British India, and what we found in the farm-houses in Holland and France, and later in the rice-godowns and tobacco-warehouses in Java, and in respect to the rat-population of ships.

In those plants or animals which form colonies from small beginnings, and thus habitually isolate themselves in smaller groups, such colonies speedily become pure for whatever genotype is given in the potential variability. Isolation is a very important factor in species-formation. A group which is completely isolated from crossing with individuals from outside must necessarily lose its variability. Darwin and Weismann have thought that, after isolation, a group could continue to change by a heightening of its variability through natural selection. The obvious reason for this idea lies in the fact, that in their time the real nature of geno-variation of its cause was not understood. The variability of an isolated group is limited, and the smaller the group, the more limited its potential variability, the sooner it will be pure altogether. If a few individuals of a variable group stock an island the population will soon be pure, and if two little colonies start on two islands, each island may have its own local and pure species after very few generations. The fact that islands are frequently found to have species of plants or animals which exist nowhere else, need not be taken as proof for the adaptation of these species to the conditions on those islands. To explain how all the individuals on one island have come to be pure for one set of characters, we need not ascribe any selection value to those characters.

Weismann has built up a complicated structure of hypotheses upon hypotheses about the mechanism of heredity. In his last writings he assumed a reciprocal relation between the determinants for an organ and this organ itself, so that, if for any reason some kind of determinant would be better nourished, they would thus become stronger and get a relatively bigger share of the available food, so that they would become still more numerous and stronger, and the organ in question would be still better developed in the off-spring of the modified individuals. This hypothesis is simply an attempt to explain the way in which modifications, the effect of the environmental factors of development could become transmittable. It is essentially Lamarck's theory. If it were true, that occasionally modifications were seen to be inherited in the way in which we see special characters inherited in higher organisms, Weismann's theory of germinal selection would become a plausible explanation, but as the facts stand, the hypothesis is not justified by them. As we will see later, the theory, that the genes are essentially chemical compounds with autokatalytic properties admits of an explanation of the similarity between properties of daughter-cells and mother-cells, as due to a quantitative preponderance of certain genes. This is a hypothesis somewhat similar to the Weismann—Lamarckian one, with the exception that it recognizes the fundamental difference between these quantitative differences and the qualitative differences which cause hereditary differences, the presence or absence of genes.

A group of organisms may become pure for a genotype which causes them to possess some organ or peculiarity, which in their present mode of life is absolutely useless. A number of individuals out of this group may some day find themselves in a position to which the same peculiarity fits them extremely well. And it is often unnecessary to assume, that adaptative characters were acquired under those conditions of life in which they confer an advantage. We must remember, that there is a perpetual broadcast distribution of seeds and young animals. The seeds of the mistletoe will only develop on branches of deciduous trees, and therefore only those seeds which happen to be deposited upon such branches will have a chance to develop. All the numerous seeds which fall to the ground or which are deposited upon branches of conifers, or fence-posts will perish. Probably the same proportion of the seed of Taxus baccata is deposited upon branches of deciduous trees as that of mistletoe. Some plants, like Sedum acre are adapted to a life on walls. We are convinced that a much greater proportion of dandelion seeds settle on top of walls than Sedum seeds. If, at any day, a group of dandelion segregates out somewhere, which is especially well adapted to a life on the top of the numerous walls of France, this species will be found growing in those situations immediately. Plants and animals are continually trying to fit themselves into all sorts of conditions, they are continually hunting for a suitable place to live, and they perish wholesale in the attempt. Dump a waggon-load of sterile sand somewhere in Northern Europe. Carex arenaria will grow on it almost directly. Sterilize a piece of moist bread, Aspergillus will be found growing on it within the week. Build a pig-sty on the heath, a thriving colony of rats will be established in a few months.

We marvel at the adaptative change of colour of the alpine hare and of the stoat with the seasons. It is more than possible that animals which are now inhabiting parts of the world where snow never falls, will react in the same way to a low temperature at the time of the autumn moult.

If we notice some plant or animal, beautifully adapted to the circumstances under which we find it living we may imagine, how it has been subjected to those same circumstances for untold generations, and how it has gradually adapted itself to those conditions. But we might be deluding ourselves. It may very well be that only after the organism became what it is now, it found the circumstances under which it now exists, in other words, it is possible that an organism finds the environment to which it happens to be adapted, rather than adapting itself to an environment. There is nothing which so binds an organism to its environment as just its ability to live in it.

We want to illustrate this point. It seems as if cultivated plants and animals are being bred more and more closely toward an ideal state of usefulness, in some fixed direction. In most cases this is true. Evidently, breeds of milk-producing cattle are bred, or rather should always be bred, with an eye to their productiveness, and they are in point of fact getting more and more adapted to their circumstances, to the economical system into which they fit. But even in the case of cultivated species, it is possible to point out instances in which particular species are now ex-ellently fitted for the use to which they are put, to the environment in which they must live, whereas we know for certain that they were not developed in these circumstances, and to fit these uses.

The Airedale terrier, formerly known as the Waterside terrier is a case in point. It was bred by rat-catchers and poachers as a useful companion, loving a fight, intelligent enough to drive a hare into the poacher's net without giving tongue, unafraid of water, and game after rats. When the craze for police-dogs began, the Airedale was one of the first to be taken up. It is now extensively used by policemen and as an ambulance-dog, and it is as well adapted for this special training as dogs which were especially bred as watch-dogs and police-dogs, such as the French de Beauce, and the German Dobermann.

Another police-dog of note is the German shepherd. This dog, as well as the Collie and other sheep-dogs, such as the Dutch and the Belgian and the Brie, was bred by shepherds, and it has been used, and is still being used extensively, to help the shepherd manage the flock, guard it, separate animals of individual owners. Young dogs learn the work directly from well-trained older dogs. Hardly any dog is as well adapted to its special work as these shepherd dogs. Now most of these dogs, especially the German and the French de Brie are being used as police-dogs and they are as well adapted to this work as any other species. That it has not the same faculties which are useful both in sheep-driving and in the hunting of criminals is shown by the remarkable fact that the Collie will make an excellent sheep-dog, but is left severely alone by the policemen.

The South-Dakota Experiment station is now breeding Persian and South-Russian sheep, which were developed as fur-bearers, but which, because of their fat tails or rumps are especially adapted to countries with a heavy snow-fall where they have to starve occasionally.

If we find two rather closely related species living in quite different environment, and which show a difference which makes them fit better each in his own mode of life, we do not know whether this anatomical or physiological difference is the result or the cause of this difference in habits or in environment. A good instance is that of the house-rat and tree-rat, both sub-species of Mus rattus. The house-rat occurs in sheltered localities, in houses, barns, ships, whereas the tree-rat lives more in the open, nests in the axils of the leaves of palm-trees, forages in the trees as well as in houses near trees. This latter rat will, if pressed, swim and dive, and it will come up again practically dry, like the Norway rat. In water the fur of the house-rat becomes soaking wet, it is not adapted to life in the open. It is very probable that the difference in hair is primary and the habit of living either indoors or outdoors is the result of this difference in texture of the hair.

There are cases in which an organism, when transported into a new environment, or employed for other uses finds itself very well adapted. In other cases some factor in theenvironment may change, and the animal may find itself able to profit by the change, to adapt itself to it. The case of the Australian Kea, a parrotwhich is killing sheep is a very striking example. Attempts to introduce new species of animals or plants often fail because local parasites find themselves beautifully adapted to a lite in which thepresenceof thenewintroduction isanimportantfactor.

Of course it is impossible to rule out of court the possibility that isolation may in some instances be the cause of speciali- zation and adaptation. However, it will be seen that isolation must be of a complete nature, and the initial adaptation must be rather great, to keep the species in the environment to which it is not yet wholly adapted, and in the second place to prevent it from dying out.

If, by a heightening of the potential variability of a species, as the result of some cross, individuals are produced, which are so constituted that at the time of reproduction they prefer an environment different from that in which the other individuals live, this fact will provide a means of isolation as effective as colonization on an island. Such a group of individuals (toads which stay on land to pair) will speedily lose its potential varia- bility. They must at the outset have a somewhat high variability for the cross which was the primary cause of the production of their new character cannot have failed to have heightened the potential variability all around. Therefore the chances are that the group will finally have a genotype different in several genes from that of the parent-species. And during this reduction of its potential variability natural selection could play a rôle. Causes heightening the potential variability of the parent-species would not affect it.

In new conditions of life, in a new environment, possibilities may be open that were not so before. An organ which may have had an important tunction, may now be useless. Therefore, if formerly only individuals having the organ in a certain state of development could live, in the new conditions toleration will be great on this point, and if genes affecting the development of the organ are recombined, the final type of the new species may have the organ greatly changed. It may now be larger or smaller or altogether rudimentary without hindering an otherwisely well-adapted species from procreating itself.

The disappearance of unused organs may in some instances be thus explained. But this is a rather complicated question.

In the first place it must be remembered that no organ is as such determined in the germ. To its development very many factors coöperate, of which some are inherited and some are not. Functions are very important developmental factors. A functionless organ may in many cases remain rudimentary for the very reason that it is without function. If we regard the heavy fore-legs, and the strong musculature of the neck of the bull-moose, we may assume that these adaptations to the great weight of the antlers are in some way inherited. But we may also try to find out whether this musculature is not to a great extent caused in every individual male-moose by the weight of the antlers. If we gradually weigh the head of a young horse with shot, until it bears the weight of a moose's antlers, it may be that the musculature of its back will closely approximate that of a bull-moose.

Carnivorous animals have a very much shorter alimentary tract than herbivores. This is an adaptation to the special diet. But feeding experiments with geese and tadpoles and ducks have shown, that the length of the alimentary tract depends very largely upon the diet of the developing individual.

In those cases where we observe an anatomical peculiarity adapted to a peculiar habit or mode of life, it is very difficult to see clearly whether the anatomical character was gradually evolved by selection to meet the requirements of the habit, or whether, reversely, the anatomical peculiarity made the habit possible. We see the giraffe with its long neck and front-legs, nicely fitted for browsing upon the foliage of trees, but it remains an open question whether by a natural selection of the individuals which could reach highest, the long neck was gradually evolved, or whether a group of very long-necked individuals found it possible to reach the leaves of trees, and thus was able to migrate into regions where short-necked animals could not live. The latter possibility appears more reasonable. The circumstance of having a long neck may well isolate a group of animals, and thus indirectly open the door for further differentiation.

In some tame animals we have good evidence for the fact that an anatomical peculiarity may be the cause for a habit. Anyone who has ever tried to teach tricks to a dog knows, that it is very easy to teach a Dachshound or an Aberdeen terrier to sit up, whereas it is very difficult to teach the trick to long-legged dogs. As in the very short-legged dogs it is the tibia and fibula which are shortened most, these animals can sit down flat on the ground and hold the body up straight. Long-legged dogs must balance themselves on their feet. Many short-legged dogs learn the trick by imitation, and it is not rare to see a Dachshound sit up to look over an obstacle or to warm itself before the fire.

For Darwin, natural selection was the only cause for specific stability If a species, by continual selection, were wrought up to the point where it as perfectly adapted to its environment, natural selection would keep it there. Darwin did not think that ever a group of organisms could be really pure, really stable, in such a way, that in new surroundings it would not be able to change and adapt itself to them It appeared to him, that only continual selection could make a species pure, and only continuous selection would keep it pure At present we know, that all closed groups of organisms, groups which are in some way protected from admixture, speedily become stable automatically. The total geno-variability of every group tends to diminish even in the absence of every selection This makes it unnecessary to assume that every quality for which we find a group stable, must be necessarily useful, and have selection-value

It is evident that no species can ever become pure for a really harmful character. But we do not need to believe that all the diverse forms of life have become as we now find them, because they have each fitted themselves to special conditions of life. The very fact, that thousands upon thousands of species exist and survive on earth, all different in hundreds of the most fundamental characters, shows, that the fitness of an organism is only a matter of finding the exact niche it can fill.

It is hard to conceive of an imaginary organism, composed of parts taken from the most diverse species, which would be incapable of existing under some set of conditions or other. In a world where an echidna, a tapeworm, an elephant, a cassowari and a bee all find their living, surely there would be place for a healthy unicorn.

There is nothing which binds an organism to its environment so much, as just its adaptation to it. Young individuals will live in the same environment as their parents only, if they are equally well-fitted to live there. If they are different, so that they do not fit there, they will happen to find another environment into which they do fit, or else they will perish, just like the individuals which are like their parents but drift into the wrong environment. Think of the series of rare lucky circumstances which the favored few tapeworms must encounter, before they are safely housed in the inside of a suitable host. At every stage the chances are one to several thousand, that their right environment will be found. In all the other cases the animal perishes. Every sort of animal or plant which is now living, is adapted to its environment, but nothing forces us to assume that the ancestral forms of any organism were adapted to the same environment.

The result of crossing, wherever the forms crossed, differ in several genes, is a great genotypic diversity, a diversity which will very often result in new characters in some of the individuals. New combinations of genes may result in new colours, new morphological characters, and new physiological characters. We saw how rats of various new colours were produced in the off-spring of hybrids between two local groups of Malayan field-rats, outwardly identical, and how waltzing rats had the same origin. We know how from the cross between two marigolds, Calendula pluviatilis and Dimorphocotheca aurantiaca, new forms originated with curious shapes and colours of flowers, and how a similar effect of crossing was found in Argemone by de Vilmorin. Crossing may be the cause of origin of individuals specially adapted to special conditions. One case is that of our sandy yellow rats, which have a colour which makes them almost invisible on sand, the exact colour of a great many desert animals.

From the experiments of Nilsson Ehle we know that from a cross between hardy species of wheat, new forms may be derived, hardier than either. Extraordinary high frost-resistance will fit wheats for high latitudes, where wheat cannot now be profitably cultivated.

The species of Argemone are rather drought-resistant. It is more than probable, that among the off-spring of de Vilmorin's hybrids there are plants, which would be able to live farther out in desert regions than either of the two parent-species. Such plants, if originated in a wild state might, if included in a number of seeds carried into the desert, found a new species, specially adapted for desert life.

Very highly specialized species, such as alpine plants and desert plants, or saline plants, or animals, always live on the edge of the region into which they can just not penetrate. They are, to borrow words from the teleologists, continually striving to establish their off-spring in places where life is as yet an impossibility, even to those organisms. If occasionally individuals are produced which can live at a still higher altitude, or with still less water, there is a chance that some day among the young animals or among the seeds that try to live outside the boundaries of the species, there will be descendants of hybrids, and that some individuals will be able to thrive there, their genotype permitting it. Thus we may picture the establishing of new sub-species in new regions, new for the group. In reality, such cases are not too rarely met with. A very typical instance is the finding of a very highly specialized Kangaroo mouse, by Grinnell, in a Californian desert.

If we are familiar with two distinct species of plants, such as Lychnis vespertina and Lychnis diurna, and we see that each prefers surroundings, which are just a little different from the optimum environment of the other, we can see that this slight difference serves to keep them distinct as species. In one type of environment, in open, dryer, grassy spots, there is a great· majority of Lychnis diurna, and occasional hybrids with ves-pertina have a poor chance of establishing themselves as a group apart in this same environment. They are too far in the minority.

If in Santpoort (Holland), a new field gets into a condition such as favours the growth of Lychnis diurna, the seeds of this species, which have been drifting into the field continually for years, will finally have a chance of developing into flowering plants. The field will be stocked from seeds of neighbouring plants, and later on, a few individuals of this field may furnish seed which succeeds in establishing itself in another field, when it has come in the right condition. In such a group of plants where cross-breeding is the rule, new types of hybrid origin will stand a poor chance of establishing themselves, unless either of two things happen, namely, unless a group of them is effective-ly isolated, or unless some new character confers a very con-siderable advantage upon the individuals presenting it.

Eventually the hybrids between vespertina and diurna will get lost into the multitude of diurnas, or into the multitude of vespertinas. One of us has observed small colonies of Lychnis in Santpoort, which were evidently of hybrid origin, such as a field in which all the plants were typical Lychnis diurna with the exception that they had the teeth of the capsule erect as in vespertina. After two years the oaks had crowded out all the Lychnis plants from this field, and it was impossible to find any similar individuals in new fields in the neighbourhood. One day, however, plants may originate which would not differ from one of these two species by having the colour, or the hairiness of the other, but in a new physiological character, a constitution which made them fit to live in marshy spots. Such plants would almost certainly find the exact environment to which they were adapted by chance, and in that environment they might live for a time. If their constitution happened to be favourable, they might even continue to exist and a new species with its own genotype and set of characters would have originated.

As Bateson pointed out, there need not be anything in the striking characters wihch distinguish diurna from vespertina, in the red colour of the flowers, in the curvature of the capsule-teeth in the absence of glands, which make the former better-fitted to a dry and sunny environment than vespertina. All these things are simply consequences of the genotype for which the plants happen to be pure.

Is it true that a new form, which happens to be far in the minority will disappear into the multitude?

Let us take case of a dioecious plant like Lychnis vespertina. We find a female plant which has a gene less than the common plants. Her daughters will be heterozygous for the gene, for all the males have it. And the children of these daughters will be pure and heterozygous in equal numbers. If every plant has one daughter, the daugther will be pure for half the number of genes for which the mother was still heterozygous. In every following generation the number of genes for which a plant is still impure, is reduced by one half. The same holds true for genes which a plant posesses more than the common run of plants of the group. Therefore we can say, that if in a population of habitually crossing plants (or animals) an individual is introduced which differs from the multitude in a number of genes, $2q$, the population will again be pure in n generations.

Let us for the moment assume that the two species of Lychnis differ in 64 genes, some present in diurna and not in vespertina, and some common to the vespertina plants only. If a hybrid is produced, it either crosses into the vespertina or into the diurna population. If there is nothing especially valuable about the characters of the hybrid, in other words if its chances for surviving are the same as for those of the individuals of the species into which the hybrid merges, in six generations the last traces of the hybridization will have disappeared. If any quality, resulting from a chance combination of genes given in the potential variability of the hybrid and its descendants confers an advantage, it may take a little longer. But only if a new character, or at least a new genotype makes a number of plants grow where they are not crossed back into the mother-species, is there any chance of perpetuation. Isolation of some sort is necessary, without isolation even selection can not work against the nivellating effect of the factors tending to reduce the potential variability. In the chapter on selection we will further discuss the possibility of change in species under selection without isolation.

The very fact that Lychnis vespertina and diurna plants are infinitely more numerous than plants of hybrid origin tends to keep the two species pure and stable. Each of them has its own potential variability, which is normally very small, which may be temporarily heightened by crossing, but always again automatically reduces itself. In a region where the two species exist at the same time, and where plants with a new genotype must occasionally be produced, there is no room for a third species to establish itself, simply because of the fact that no group of plants can effectively become isolated from the "swamping" multitude of individuals of the two species.

If a new form is prevented by any circumstance from interbreeding with the parent-species, its chances to persist are very much greater. And it is evident that these chances are better in those casses, where the new group fills a niche different from that of the parent-species. For in that case it does not need to be counted as a minority in a mixture of two forms, subjected to the same factors which govern the number of individuals having off-spring in every generation.

If an individual or a few individuals found a colony, stock a field, the colony may become numerically of sufficient import-ance to persist. In groups of plants where selff-ertilization or asexual propagation is the rule, but where nevertheless occa-sional crossing is not excluded absolutely, numerous new spe-cies, of different biotypes may coexist in the same ecological niche. In such cases we understand that continually some of these species fade into insignificance and disappear, whereas new species are continually produced. We may feel sure that wherever we meet with such a group of species, such a poly-morphic group, we will find that either asexual multiplication or self-fertilization is the common mode of reproduction. Wheat, barley, Viola tricolor are typical examples of self-fertilizing poli-morphic plants, and the dandelion and stinging-nettle are good examples of polymorphic groups of plants where asexual mul-tiplication is common. Real polymorphy, the exï'stence of nu-merous different relatively pure species in the same ecological niche is only possible in such plants. Polymorphy in animals is possible only under the same conditions, in uni-cellular organ-isms, in vegetatively reproducing animals, and in instances of isolation within the species through sexual selection. The cases of the persistence of black individuals in moths and certain birds, pointed out by Bateson in his book on the problems of Genetics may fall into this category.

We have seen how it is possible, that occasionally individ-uals are produced in populations of allogamous animals and plants, which differ sufficiently physiologically to come to fit a different environment, in which they would tend to be a new species.

This mode of origin of a species must be common to the free-crossing and the self-fertilizing organisms. We may call it phys-iological isolation. Any kind of isolation must tend to species-formation, to the production of new groups, having their own

"centre of stability". This is Wagner's argument. Geographic
barriers may be effective means of species formation. A group
of animals on an island, or on the other side of a wide stream
must have a potential variability smaller than that of the multitude of its species. We do not need to assume that a species
gradually, by a slow process of natural selection became adapted to very special conditions, but that the peculiarity which
makes the group fit may have been the "accidental" result of a
recombination of genes, and the cause which drove the first individuals of this sort to establish themselves in the new conditions or perish. A species may be pure, without potential variability, before some of its members colonize, in such a case no
new species will be produced. But if it has some potential variability left, or if its variability happens to be heightened by a
recent cross, a colony derived from it may soon have its own
genotype, and therefore its own characters, its own species
type. Each little American desert has its own species of
ground-squirell, which is relatively pure, and somewhat different from all the other species.

Of course species-differentiation is possible in allogamous organisms in cases, where barriers of any kind are absent. In our
hypothetical examples we assumed, that there was a considerable interchange of individuals going on throughout the whole
range covered by the group. If the conditions are such that a
gene, introduced at one end of the range may be transmitted in
a few dozen generations to individuals at the other end of the
range, the whole group will be one species, it will tend all the
time to reduce its potential variability, it will tend to assume
one single genotype, and consequently one single phenotype.
Those plants and animals will not easily form local species, local forms. Examples are the English sparrow, the Norway rat.
But we can see that in a slow-moving animal the chance for a
gene, introduced somewhere by crossing to become part of the
eventual genotype of the group is very much greater than in
roving animals, with a rapid dispersal

If every individual has on the average only one descendant,

we took it for granted that an individual, heterozygous for one gene would surely mate with a homozygote, and on this assumption our calculation of the chances against the persistence of varietal characters was based. But it will have to be conceded, that an important factor in this chance for persistence is the rate of dispersal of the organisms. If we are dealing with organisms which move about very little, and whose young or eggs or seeds are not commonly transported over great distances, we must recognize that the chance for a varietal difference, depending upon a genotypic difference in one or two genes, to persist in the absence of any selection is materially heightened. In self-fertilizing organisms varieties are incipient species, each variety is fully protected from the random crossing with the great majority of individuals of the type of the species. And now we must admit that in allogamous organisms, this same chance for a slight genotypic difference to persist within a species, varies from practically zero in rapidly spreading organisms to something which approaches more or less to the same chance in self-fertilized organisms. It is evident that in both instances the automatic reduction of the potential variability of any group works against the persistence of varietal characters, in a measure proportionate to the percentage of individuals procreating out of total number produced.

In animals or plants which hardly move about, the chance for inbreeding, and therefore the chance for a mating between individuals with the same genotypical peculiarity becomes greater, and it is certainly not infinitely small in extreme instances. Snails must for this reason be very much more liable to produce local species than falcons, even if a group of snails covers the same territory as a group of falcons and in the absence of barriers, there will be a chance for the snails to produce local species out of its varieties, but none for the falcons. The clearest instances of circumscribed groups of organisms within which free-crossing is the rule, are the cases in which such groups are locked up into a circumscribed area in the way in which aquatic organisms are confined in a lake. Here we may limit our comparison to allogamous organisms. We will probably find that even in a lake of a few acres there are local forms of such animals as fresh-water snails in different localities of the pond. But we will find only one species of trout in it, and not slightly longer animals along one bank and slightly higher along the opposite bank.

In animals with sedentary habits, we are likely to find this existence of local forms. If the difference between such groups is due among other genes to one which has a very decided influence upon the appearance of the animals, it may be possible to trace the territory covered by individuals carrying this gene. And we will probably find that different genes have different and over-lapping territories. As, however, the action of most of the individual genes is slight, and as very often several genes may influence the development in analogous ways, the effect will be, that from one end of the range to the other we will see the animals of such a group gradually change from one local form into the other. Such cases are the despair of the systematic zoologist. For purposes of classification he will have to choose between lumping all the forms into one species, or selecting some of the most striking types as specific types.

The facts in these cases are very much obscured by the circumstance that among the causes of this multiform appearance of such a group, which is something very different from polymorphy, there may be two things, first toleration of chance combination of genes, and secondly real adaptation to the environment. When we observe a series of birds of one group, and see that the individuals from the coastal marshes are darker, and give place more inland to lighter and lighter birds, up to very pale desert-birds, part of this diversity may be real adaptation. At the same time a corresponding diversity in the number of scales on the tarsus may be a case of mere chance, plain toleration. And both sets of characters may be mainly due to genotypic constitution, and be found to breed true in any environment.

From the work of the Berkeley museum of Vertebrate Zoölogy we could quote several instances of gradual change of characters throughout the range of a "Paarungsgenossenschaft" of animals, the case of the gopher, of the field-mice, of the garter-snakes, fox-sparrows. It is significant that animals with greater power of dispersal are more apt to be monotypic throughout the whole of the area into which they are confined through reasons of ecology. Jack-rabbits and coyotes are less liable to produce several local forms within a territory without barriers than the slower moving and less roving animals.

Where the "type-specimens" are really commoner than "intermediary stages" these types may be called species in good accordance to my definition of the term. In practice it is believed that a new species upon which a new specific name is confered, is a stable group, somewhat variable perhaps, but centering about a mass of individuals all like the type-specimen, and thought to have a vaguely appreciated tendency to remain in this condition. Our definition of species, as groups of organisms, so constituted and situated, that they tend, under conditions, which promise to be permanent, to reduce automatically their potential variability also defines the species of the taxonomist.

MUTATION.

DARWIN believed, that occasionally individuals are produced which differ from their ancestors in a marked way, "sports," but he does not seem to think that such sports have ever played an important part in evolution.

De Vries states as his opinion, that new species come into existence, not by selection after continuous geno-variation, but suddenly, as the result of abrupt changes of the "inheritable". According to him, new species differ from parent-species in several characters, which they have acquired with one stroke. He further believes that species are on the whole stable, only occasionally breaking out into periods of mutability.

When judging the work of de Vries, it must be borne in mind that these ideas were set forth just before Mendel's work was rediscovered and that Genetics in these days had no connection with Biomechanics. It was possible to believe that the inherited consisted of numerous determinants, which mysteriously called forth definite kinds of organs or characters in a direct way. We see now, that the genes act upon the final characters of an organism by influencing its development at some stage or other, and we understand, that there need not be any relation between the number of new genes acquired or lost or changed and the importance of the changes effected in the characters of the organisms in which these changes take place.

We know, that a gene may have an important influence on the development of a given biotype, whereas the same gene may not affect the development of another group of organisms at all.

Neither the periodicity of supposed mutations, nor the idea that a spontaneous change of the genotype must necessarily translate itself in a discontinuous variation, are essential to the theory, that such abrupt changes may occur and may play a rôle in evolution. We know, that the effect of a great number of genes studied in the most diverse organisms, is very slight. We do not think any author on the subject has accepted the idea of a periodicity in mutation.

If genes are lost or acquired spontaneously at times, such an occurrence may very well pass unnoticed, if the gene in question does not happen to be an obvious factor in the development of organisms with this genotype.

As, theoretically, a mutation can consist of either the spontaneous acquisition of a gene, or of the spontaneous loss, we are confronted with a difficulty. Is it possible in a given instance to know whether a gene has been added or whether a gene has been lost? The point is rather important, because of the obviously unequal interest attached to these two theoretical processes. If we see two organisms who obviously differ in one gene, how can we judge which one has the gene in question and which one lacks it? We think that dominance of any quality of an individual over a corresponding character of another, proves the presence in the first one of something which is absent from the other. If two animals differ in colour because from the germ of one of them there is lacking an indispensable link in the chain of factors necessary for pigmentation, the hybrid will be coloured because it inherited that necessary factor, if only from one parent. In such a case we assume that an individual which is impure, heterozygous for a gene will show in its development the action of this gene as strongly or nearly as strongly as an individual which is pure for it, in other words that a single dose of the gene ultimately has approximately the same effect on the development as a double dose. This is the "presence and absence" theory as I proposed it in 1908.

We have to go carefully here, so as not to be deceived by the apparent fitness of the hypothesis, for, as Shull has pointed out, a hybrid from a cross between individuals of which one

had and one lacked a certain gene, might be like the parent
lacking it. Namely, if the gene and the developmental-process
influenced by it were of such a nature, that one single dose
would not affect the development, whereas two doses would.
Theoretically such a case might well be imaginable. Shull has
invented a very elegant chemical model to illustrate the process. If we represent a gene by an acid, and the organism to be
affected by an alkaline solution of lithmus, we may so choose
the strength of the acid and of the solution, that the latter will
turn red by adding a certain dose of the acid, but will remain
blue if we add only half the amount. A priori, we cannot exclude the possibility of such a process, because we do not know
enough about the real nature of the genes, and their action
upon the development. But there are nevertheless indications
which show that in reality dominance means presence of a
gene, and recessiveness means absence.

If it were true, that there are cases in which a gene, inherited in only one gamete has not sufficient influence to modify
the development of the organism, and to alter its characters,
but in which the same gene, inherited in both germ-cells will
visibly affect a certain character, this case would constitute
one extreme instance of a whole scale of possible intermediate
instances, ranging from cases in which heterozygotes are fully
as much changed by two doses of a gene as by one dose, down
through cases in which heterozygotes showed only 90 %, and
70 % and 50 %, 20 % and 10 % of the action of a gene, as
compared to the action of the same gene present in both gametes of the zygote.

If it were true that we could put together a list of instances,
in which heterozygotes showed the action of a gene inherited in
one gamete, to be only 90 %, 70 %, 60 %, 50 % of that of the
same gene inherited from both parents, the case would be
undecided. If, for the sake of simplicity we call the characters
influenced by presence or absence of a gene black and white,
and we found that a hybrid between a black and a white, differing in only one gene, were grey, midway between the ex

tremes, we would not know whether this grey should be interpreted as 50 % black, due to the presence of only one dose of a black-making gene, or as 50 % white, due to one dose of a white-making gene, which the white parent had more than the black one. And a case in which the hybrid between a black and a white were 70 % black, might also be interpreted as 30 % white, caused by insufficient action of one dose of a white making gene. There would be no way to decide which of the crossed individuals, the black or the white, had a gene more than the other. Now the fact is, that, whereas we know a few instances in which a heterozygote is just noticeably distinguishable from a homozygote, there are no cases in which dominance in a heterozygote is about 80 % or 60 % only, and in which we can be sure that there are not more than one gene responsible for the difference. For here lies the difficulty. If we observe a mulatto who is 60 % as black as his negro father, we do not know whether this father differed from the white mother only in one black-making gene. We know, to confine ourselves to colour, that there are genes which tend to make pigmentation deeper, and that there are others which make it lighter, so that if two animals or plants are crossed, of which one has a darkening gene and the other a lightening one, and each lacks the other, the colour of the hybrid may be intermediate as a result of the fact that the two genes counteract each other's effect upon the development. A few instances: Black mice have a gene more than certain yellows, but they lack a certain other gene which agoutis have. The same relation is found in rabbits. Yellow agouti mice and rabbits have this last gene, but they lack the one which blacks have more than yellows. If we now mate a black to a yellow agouti, the hybrids will be agouti, which is as exactly intermediate between the colours of the parents as one could only wish. Bateson found that the Leghorns have a gene, which tends to reduce the pigmentation of the skin. The black-skinned Silkies lack this gene, but possess another, which makes possible an intense blue-black pigmentation of the skin. Hybrids between the two species have both genes, and as a result their skin is pigmented, but very much lighter than that of the pure Silky. If, however, the Silky is mated to light-skinned fowls lacking the "lightening" gene, the hybrids are as intensely pigmented as the real Silkies.

Now it would seem, as if it would always be easy to distinguish between cases in which an intermediate hybrid were half-way between his parents in some character because he inherited from each a gene, which genes counteracted each others effects, or whether the intermediate condition of the character was due to the fact, that one single gene if transmitted only in one gamete were less effective than in double dose. For the children of such a hybrid would in one case show a bi-factorial segregation, $9 : 3 : 3 : 1$ and the off-spring in the other case would segregate into a $1 : 2 : 1$ ratio. But we must remember that in a few instances we get an apparent monofactorial segregation, $1 : 2 : 1$, in those instances which Bateson has designated by the name "spurious allelomorphism". An individual, impure for two genes, sometimes produces only two kinds of germ-cells, instead of the ordinary four, AB, Ab, aB and ab, namely only germ-cells containing either one or the other gene, and none of the other categories. Now such instances are easily mistaken for cases, in which heterozygotes show the action of a gene, which they have inherited in one germ-cell only to about half the extent of what it amounts to in homozygotes. One of us has investigated a case in mice, in which such a spurious allelomorphism existed between the gene which agouti animals have more than black ones and that which pigmented animals have more than albinos. Agouti hybrids in this series, produced by mating an albino to a black, did not produce nine agouti to three black and four albinos, of which agoutis there would be several kinds, including homozygotes, and of which albinos there would be two kinds, but two agouti heterozygotes, to one black and one albino homozygotes. If the two factors concerned had not been known beforehand, in other series of experiments, in which they were transmitted independently, it would have been presumed that the difference between the albino and the black was a difference produced by presence and absence of one single gene, and that the agoutis were the individuals heterozygous for this gene, especially as they might be termed intermediate.

Those cases, where hybrids have been found to be intermediate between two parents, and which look as if only one gene made the difference between these parents, in as far as the character studied was concerned, are few indeed. One of the best studied cases is the almost classical instance of the Blue Andalusian fowl. In some species of fowls three colours exist, black, blue and white. In most of these, all the three colours are bred and can be shown at poultry shows. This is the case with black, blue and white Wyandottes, and Orpingtons, the Dutch species of Witkuiven, sometimes called Polish, and the Kraaikoppen. In some species however, as in the Andalusian, and in one of the fighting breeds, only blues are wanted, and shown. In most cases studied, we find that blues are produced by crossing a black and a white, so long as the cross is confined to differently coloured members of the same breed. Such blues, if mated with each other produce about 50 % blues and the remaining chicks will be blacks and whites, more or less splashed. Blue bred to white gives as many blues as whites, and blue to black produces as many blues as blacks. Both the whites and the blacks breed true. This all works out well on the hypothesis that the difference between black and white in these breeds is caused by presence or absence of one single gene, and we can either assume that this gene is a white-making gene absent from black, or a black-making gene absent from the whites, so long as we assume that heterozygotes, individuals having inherited it from only one parent show its action only to a limited extent and are therefore only 50 % white, or 50 % black, which amounts to the same thing. Punnett in his interpretation of breeding-experiments with Andalusian fowls together with Bateson, choose the hypothesis that the blacks

10

have a gene more than the whites, and that the blues have this colour because they are only half-black.

We must remember that there is an alternative hypothesis, namely that the blacks have a gene more than the whites, and the whites another gene more than the blacks, and that the blues are blue because they inherit both genes, and the black-making tendency of one gene counteracts the white-making tendency of the other. This would imply that the rôle of the white parent was not a passive one, but that the white is a dominant white. It would permit the explanation of the existence of true-breeding blue fowls. In the case of the Andalusian blue, and the Kraaikop, however, where there are no true-breeding blues, we need the secondary hypothesis that there is a mutual repulsion between the gene which the blacks have more than the whites and that which the whites have more than the blacks.

Is there any way of choosing between the two explanations? We think that there are some facts which make us prefer the hypothesis, that in these cases we are concerned with two genes, which where they are together, counteract each other's effect upon colour. In the first place, as we saw, the theory that the blues are blue because they are heterozygous for a factor for which the blacks are pure, but which is absent from whites, implies that the white is a recessive white, and contributes nothing. If this were true, we could substitute any other recessive white fowl in the cross-blue Andalusian white, for the Andalusian white, and we would obtain the same results. Now, when in 1911 we mated a blue Andalusian male to a recessive white (Wyandotte) hen, we did not obtain blues and whites in equal proportion, as we would have obtained if the hen had been a white Andalusian. (Fig. 17). Half of the number of young were black (12), and half of them were blue (13). We can explain this by saying that the white Wyandotte hen, but for lack of pigment would have been black. But then it becomes clear that we should expect only black young, unless the male, who was blue, transmitted something to half the number of his germ-cells, which made the chicks growing from them blue. In other words, this result of the Andalusian-Wyandotte cross tends to show, that blue is dominant to black, in other words that the blues have something more than the blacks. Now there is another fact which proves this. It is well known that in chickens there have been noted several instances where females, heterozygous for a gene produce sons with it and daughters without it, provided her mate does not pos-

Fig. 17.
Results of crossing a male blue Andalusian to female White Wyandotte.

sess this same gene. A Silver-Wyandotte hen, or Silver Sebright hen or Assendelver, if mated to a Golden cock will produce golden pullets and silver cockerels. There seems to be a mutual repulsion between the gene which is responsible for female sex and a few other genes. Four or five instances have been noted in the fowl. In all these instances we can see this sex-linked colour inheritance only in cases, where a female of the dominant colour is mated to a male of the recessive colour. It therefore becomes significant that according to a breeder of

Witkuiven in Holland, Mr. Smits, there exists such a case of
mutual repulsion in this fowl. Blue females, mated to black
males give as many black as blue chicks, but the blues are all
females and the blacks are all males. This can be explained
only on the assumption that blue is dominant over black, in
other words, that the blue colour here is due to the presence of
something in addition to the set of genes of the black animals,
something which is derived from the white parent. (Fig. 18).

Fig. 18.
Sex-limited colour-inheritance in Witkuiven (Polish).

We do not think that in the light of the result of further
breeding-work with Andalusians, especially of crossing-experiments of Andalusians with other breeds, the hypothesis that
the three colours are due to homozygosis, heterozygosis and
absence of one single gene can be upheld. The dominance of
blue over black, which is evident from the sex-linked colour-inheritance in Witkuiven further proves, that we are dealing
with two opposing factors.

There are cases on record it is true, where heterozygotes
could be distinguished from homozygotes. Nilsson Ehle's

example of the black colour in oats is a very striking one. But we have tried to show that as there is no series of cases of incomplete dominance, ranging from nearly complete dominance down to 50% dominance, we may feel safe in assuming that if two forms crossed differ in only one gene, that parent, which has the dominant character, had a gene more than the other. For the present we think, dominance is a good criterion for presence of an additional gene.

We think it will be necessary in this connection briefly to treat of a few cases, which have been held to show how the presence of a gene can be recessive to its absence, namely those cases from which it is said that they show, how a character can be dominant in one sex and recessive in the other sex. A typical example is the case of the inheritance of horns in sheep, brought forward by Woods. (Fig. 19).

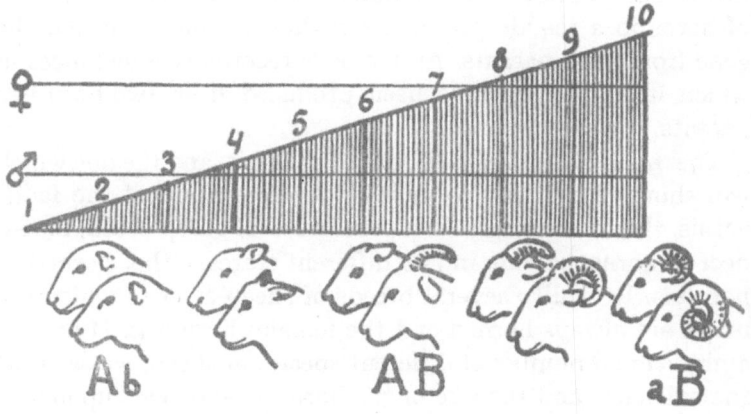

Fig. 19.
Diagram to illustrate the effect of different combinations of genes influencing horn-development if tendency to horn-formation is unequal in the two sexes.

Woods mated Dorset to Suffolk sheep. Dorset is a species of domestic sheep in which both sexes are horned, and in the other species both rams and ewes are horn-less. Woods observed that the male hybrids were horned, and that the females were hornless, like the Suffolk. If a F2 generation was raised, by mating the hybrids inter se, it was found that of the rams, three times as many were horned as horn-less, and that among the ewes there were three horn-less to one horned. Woods offered the following explanation. He assumed the existence of one definite horn-determining gene in the Dorset breed. All the animals of the Dorset breed are homozygous for this gene, and all the Suffolk sheep lack it. The hybrids are heterozygous for it, and the gene is fully dominant in males, even if present in single dose, but a single dose is without effect on females, and they have to be homozygous for it to have horns. In other words, presence of this gene is dominant in males, but recessive in females. Now, so far as the facts have been stated, they are fitted by the hypothesis. But this hypothesis implies among other things, that a male can only be horn-less if it lacks the gene, so that even with heterozygous horn-less females, a horn-less father can never have a daughter with horns, as presence of horns in a female pre-supposes that she has inherited the gene from both parents. And Woods records two instances in which horned ewes have been produced from two horn-less parents.

The main objection to Woods hypothesis, and the one which will show us the way to another interpretation of the facts, is this, that according to it a pure breed of sheep could, in respect to horns, be only in two different states, either horned or hornless. In reality several breeds of sheep exist in which the males are always horned and the females horn-less. If we examine a great number of different species of sheep, we see that the existence and the size of the horns must depend upon several different factors. There exist breeds in which both sexes are truly horn-less, as the Suffolk, the Texel, and the Lincoln. Next come the breeds in which the males have small scurs or very small horns occasionally, but the females never. The Cheviot breed is a fair example. In the breeds in which the females generally have small horny scurs, the horns of the males are much larger than in the breeds where the females are truly horn-less. In those species where both sexes are horned there is always a very great difference between the size of the horns in the two sexes. (Heather sheep). In those breeds where the females have large horns those of the rams are enormous — Dorset, Merino, Algerian.

In all the breeds the males have a greater tendency to horn-formation than the females, and horn-formation and shape of the horn is much influenced by early castration. Animals of one genotype differ in the amount of horn-formation according to their sex. There is no evidence for the assumption, that all those genes which must influence the growth of the horns, must by their coöperation to the development tend to make the horns grow longer. As we know that absence of horns is more or less dominant in cattle and in goats, we are safe in assuming that there can be genes whose coöperation to the development tends to make the horns stay shorter, as well as genes whose action results in a lengthening of the horns.

Not to make the case too complicated, we may imagine a wild breed of sheep, possessing two genes, influencing horn-formation, one tending to reduce the size of the horns and one tending to increase the size. We will call these genes A and B, and we imagine the sheep to be all $AABB$. The sheep are in an intermediate position in regard to horns, and the males are horned and the females horn-less. If now we imagine, that two different breeds exist, differing from this wild species, one by absence of A and one by absence of B, the result will be that one of these breeds will be of the composition $aaBB$, and the other will be $AAbb$. Therefore the horns of the first breed come under the action of the factor B, promoting horn-growth, and in this breed both sexes are horned, although the males will of course have horns longer than the females. The second breed will lack B, and at the same time it has A, which has a reducing effect. This breed will be wholly hornless. If ever we cross the two breeds, $AAbb$ with $aaBB$, the hybrids will have both A and B, and even if we assume that as heterozygotes they show the effect of a presence of either factor as well as homozygotes, we will see that they will in respect to horn-form-

ation ressemble the wild species $AABB$. In other words the
male will have horns and the females will be hornless. Matings
of such $AaBb$ hybrids inter se will give the customary $9\ AB$,
$3Ab$, $3aB$ and $1ab$ individuals. For males this means: $9AB$,
horned, like the wild, $3aB$ horned, like the long-horned breed,
$3Ab$ horn-less, like the rams of the horn-less breed and $6ab$,
probably horned For females the result will be $9AB$, horn-
less as the wild breed, $3Ab$ horn-less as the horn-less breed,
$3aB$ horned like the forned breed and $1ab$ probably horn-less.
In all for males, 12 horned t 4 horn-less, which is $3:1$ and for
females 12 horn-less to 4 horned, which is also $3:1$. This tallies
very well with the facts as observed by Woods. It is preferable
in three points, firstly because it does not need the additional
hypothesis that heterozygotes are markedly different from
homozygotes, secondly because it accords better with the
existence of pure breeds with all stages of horn-development,
and thirdly, because it explains the production of horned
daughters by horn-less parents, e. g. the production of horned
$aaBb$ females by an $Aabb$ father and $AaBB$ mother, both
horn-less.

We may not forget that, at the time when Woods published
his results, it was still rather universally believed that the genes
were things which each "determined" a single organ or a
single quality of an organism. It is obvious, that the work was
begun with the assumption, that he was dealing with a single
"unit-factor," determining horns. And it is probably for the
same reason, that Woods assumed that the horned breed
should have a gene more than the hornless, rather than reversely, notwithstanding the fact that he was perfectly free
to choose either hypothesis.

A series of genotypically different families, which each has
a certain character developed to an extent of its own, but
in which in every strain the males differ materially from the
females such as we encountered in the sheep, must be of rather
common occurrence. Another good example is that of the
Silver Wyandotte. In this species the males are very much

lighter coloured than the females. The ideal animal is uniformly marked all over, and every feather must be white, set off with a narrow, but uninterrupted black border. In animals which are too dark, the black border encroaches upon the white centre, and in extreme instances the feathers become black with a narrow whit ish streak along the shaft. On the other hand, in animals which are too light, the feathers may become almost white, with a small black tip. The standard of perfection of this breed of fowls calls for a marking of the feathers equal in both sexes. Now it has been found in practice, that it is impossible to establish a family of Silver Wyandottes, in which perfectly marked cocks are regularly produced, as well as hens which come up to the standard of perfection. As the males have a tendency to be lighter marked than the females, the SiverWyandotte as it is exhibited in poultry shows, consists of two distinct species, a male-producing strain, and a strain which furnishes exhibition hens. Some fanciers breed both species, and others specialize in one of them, and strive either to produce prize-winning cocks or perfect hens. The cocks, in the species which produces well-marked hens, are very light, and reversely, the hens in the strain which is kept for the production of good males, are too black to be able to compete in the shows. Here the number of genes which influence the shade of the birds must be larger than one, for there are families of Silver Wyandottes, of which the members are far lighter than even the hen-breeding strain, so light, that the cocks are whit ish and the hens even are much too light. And also, there are strains of Silver Wyandottes of which even the males are too black. The difference in shade between the sexes in every family must ultimately depend upon the genotypic difference, which makes some animals be hens and others males. In many species of chickens a similar difference exists, e. g. in Barred Plymouth Rocks.

If it were true that the loss of a gene might result in the production of a new form, dominant in the quality affected by this gene, over the form from which it was derived, it would be quite impossible to know whether a new dominant form, spontaneously originated, thanked its origin to the acquisition of a new gene or to the loss of one which had been present. But we have seen, that we have no reason to assume that the loss of a gene could produce a form, dominant over that from which it was derived. And this is further strengthened by the observation, that as yet the only mutations which have been verified in a sufficient way, have, with a few significant exceptions, always resulted in the production of a recessive form. As I will later show, we may not conclude that new genes have been acquired spontaneously in a species, because we know of the existence of tame cultivated individuals, which possess genes which the wild parent-form has not. In this connection it is * very significant to observe the fact, to which Bateson draws attention in his "Problems of Genetics," namely that plants which, as the Sweet-pea, cannot possibly be crossed with other wild species, and which have given rise to hundreds upon hundreds of distinct domestic species, have not produced dominant forms.

Lotsy goes so far as to deny that spontaneous change of genotype, real cases of mutation, exist. It is true that it is very difficult to obtain satisfactory proof, that a real mutation is witnessed. We know of no case in the literature on the subject, in which it was probable that a new gene was spontaneously acquired. We may therefore simply ask the question: What is necessary to prove that the production of a new form is due to a really spontaneous loss of one or more genes? It is obvious that the occurrence, which can most easily be mistaken for mutation, is the production of an individual lacking a gene, from a parent, or a pair of parents, impure, heterozygous, for the gene in question. When a new form is found wild, and it is found that individuals of this new form breed true, it should not be allowable to speak of mutation. Formerly this mistake was made more often than at present, for instance by Blaringhem. It must here be stated, that it has unhappily become custumary for some American authors to use the term mutation as synonymous with biotype. They call "mutation" each individual with a new genotype, no matter how produced, for instance they write about the selection of the best mutants in the descendants of a cross. It is very probable that this confusion is primarily due to the fact that de Vries, in reviewing the plant-breeding work of Mr. Burbank, assumed that very many of Burbank's novelties had originated by mutation, notwithstanding the fact that on Burbank's evidence they had originated as descendants of hybrids.

The fact, that a family of animals suddenly produces a new recessive form, although it has never before during a long series of controlled generations produced this novelty, may not be spoken of as mutation without further proof. We know that during all the time that this recessive form has not been produced, a number of the animals may well have been heterozygous for the distinguishing gene, and the chance mating of two heterozygotes may be responsible for the birth of the new recessive form. To be sure of the occurence of a mutation in animals, it is necessary to show that an individual which is proved to be homozygous for a certain gene, has nevertheless produced at least one germ-cell which did not contain it. We have done this ourselves in mice.

In one case we were dealing with a family of animals which all possessed the gene which we called G, and which was known to produce the difference between black and agouti. The animals were strictly inbred. I every generation brothers were mated to sisters. This circumstance made it subsequently possible to make probable, that a real spontaneous loss of a gene was the cause of the production of the animals without G.

We found, that the two parents of the mutants were both heterozygous for G, and that of the two grand-parents, one was pure, GG, and the other impure for this factor, Gg. The great-grand-parents, when tested, proved to be both homozygous for G, GG. The male gave 34 young, all having G, when mated to gg females, and the female gave 27 young with G, when mated to males without G. This proved, that both these animals were homozygous for G, whereas at least one of them must at some time have produced at least one gamete from which G was lacking. In another instance two mice were tested by mating them to animals lacking in the factor F, a gene which fully-coloured mice have, and silvered lack. The male in these test-matings gave 24 young with F, the female gave 12 young with F, and none without, when their respective mates were lacking in F. Therefore these two animals must have been both homozygous for F, FF. Nevertheless we found that one of their young, when they were bred together, a male, was heterozygous for this same factor Ff. So here again one of the homozygous, FF animals must have given off at least one gamete, in which this gene was not present. To prove a real case of mutation in animals, it is absolutely necessary to show that an individual, which by suitable test-matings has been proved to be pure for a certain gene, nevertheless produces a germ-cell without it. It might be thought, that strict inbreeding would necessarily show whether in a given group of animals all were pure for a certain gene. But we will see, that the production of a new recessive form can be the result of a perfectly normal Mendelian segregation and still take a number of generations to realize. It even sometimes happens, that one half of the number of the individuals of a group are impure for a gene, which has a marked influence on animals of their biotype, without the production of the corresponding recessive type. These cases are those, in which for any reason two gametes without the gene can not combine. In a number of animals, it has been noted, that a mutual repulsion exists between a certain gene and that which females have more than males. If in such a case a female is impure for the gene in question, whereas the male is pure, only two kinds of off-spring are produced, females heterozygous for the gene, and males pure for it, because all those germ-cells, into which the sex-determining gene enters, and which will ultimately give females, will lack the gene. A male can never be without this gene, or even impure for it, and therefore this peculiar state of things is permanent in the family. By observing several pure-bred generations of such material, the peculiar fact that all the females are heterozygous, cannot be brought to light. A breeder of Silver Wyandottes has no idea of the fact, that all his hens are heterozygous for the gene which distinguishes his breed from Golden Wyandottes. He does not know, that his silver hens produce as many germ-cells from which this gene is lacking, as eggs having it. And at first sight it looks strange that all females of at least two different turtle-doves, and all females of the domestic pigeon should be heterozygous for a gene, indispensable for pigment-formation, and that their genetic constitution is such a one, as we would expect hybrids with one albino parent to have. And yet, if we mate any silver-coloured hen to a golden cock, we find that only fifty per cent of the off-spring are silver, namely the sons: the daughters, being gold-coloured. This curious inheritence was first brought to the notice of one of us some fifteen years ago, when visiting the town of Assendelft in Holland. In the neighbourhood of Assendelft and Landsmeer, the raising of ducks and fowls for eggs is an important industry. The common fowl in those regions is called Assendelver, and is practically identical with the Gold and Silver-pencilled Hamburg breed. For the production of pullets that are destined to be kept for egg-production, the farmers mate silver hens to golden cocks. The resulting chicks come in two colours, white and yellow. Only the yellows, which will ultimately prove to be all hens, are raised, and the whites, which are cocks, are said to be destroyed. Later, one of us paid a special visit to the neighbourhood to collect some data, and found fourteen cases in which chicks were raised from the combination of silver hens and golden cocks. 249 of these were silver, and all males, and 243 were golden and pullets. He found nine broods of chickens from two golden parents, and all the 104 were golden.

The other cross, golden hens and silver cocks produces chickens of both colours in both sexes. He observed fifteen lots of chicks thus produced. They comprised *162* Silver males, *163* silver hens, *165* golden cocks and *160* golden hens. In most of the farms nothing but golden chicks were kept, and these were all pullets. He was told that the cockerels were weeded out in the first week. When living near Paris we happened to find out what happened to these cockerels. Every spring great consignments of very young chicks arrive at the Halles there from Holland. They find a ready market as baby chicks, but never a single pullet is found among them.

If we mate a female dove to an albino male, we find that all the daughters are albinos and that the males are coloured. The same state of things has been found by several authors in quite a number of instances. From personal observation we know that the following genes may show this repulsion from the "female-determining" gene: That which Silver Wyandottes and Silver Assendelver fowls have more than Golden-coloured ones, that which Duck-wing bantams have more than Black-reds, that which Brown-red English fighting bantams have more than Black-reds. In pigeons one of us found one of the genes necessary for pigmentation in the wild European turtle-dove to be in this condition. By others a long list of genes have been found to behave in this way. Such are the gene which barred chickens have more than blacks, that which green canaries have but brown ones lack, and the gene which by its presence or absence makes the difference between normal Abraxas grossulariata and its variety lacticolor. That ordinarily a family of Silver Wyandottes, or of Barred Plymouth rocks breeds true, must be due to this curious repulsion. On the other hand, we observe that if at some day a pure-bred family of these fowls does produce a Gold-coloured or a black hen, we may not call such an occurence by the name of mutation, because we do not at all know what cause lies at the base of this repulsion. From our experiments with mice we know, that in some cases the genes G and A are inherited quite independently one from the other, and do not influence each other's distribution over the gametes produced, whereas in other cases they show a mutual repulsion. This shows that such a mutual repulsion is not given by the nature of the genes themselves, but by some cause outside them. When anything happens which disturbs this repulsion, a male-giving egg may be produced without this gene, and with the birth of a male heterozygote, individuals of the new recessive form must necessarily be produced, as all the females are heterozygous.

No amount of inbreeding, no long series of generations during which a family of animals has been purely bred, suffice at all to show, that the production of an individual with a new recessive character is caused by a mutation. On the other hand, the fact that at least two cases have been observed in animals of the spontaneous loss of a gene from a gamete produced by a homozygote, make it probable that real mutation, real spontaneous loss of a gene is a phenomenon, which we can hope to observe now and then in favorable circumstances.

Some authors seem to think, that in plants it is always easy to distinguish cases of real mutation from instances in which a recessive is produced by a heterozygote. But we must remember that, just as we have a series of instances in animals in which a family may contain heterozygotes, without producing the corresponding recessives, we know a series of instances in plants in which individuals may be heterozygous for even a number of factors without the production of the corresponding recessives. These are those instances in which ordinary monocious plants seem to consist not of one individual but of two individuals rolled into one, which two may each have their own genotype.

In such cases the pollen may be genetically different from the ovules, so that in crossing experiments the same plant may prove to be very different genotypically according to whether we use it as the male or as the female parent. The best-studied examples are to be found in the carefully planned work of Miss Saunders with Matthiolla. Other examples, which are important for an insight into the mutation question are found in the work of de Vries with Oenothera. De Vries studied crosses between Oenothera biennis and O. muricata, between O. Biennis and O. Hookeri and several similar species, the results being sufficently similar to admit of taking one cross as a typical example for the purpose of this discussion. The peculiar type of inheritance first discovered by de Vries in these hybrids was confirmed by several authors, e.g. the cross Biennis × muricara was repeated by Davis, with the same results as obtained by de Vries.

The hybrids which have Biennis as the mother, differ from those which have it as the father. Cases of unlikeness of reciprocal hybrids are rare and they show, that the germ-cells given off by a certain species as the father are not genotypically identical with those produced by that same species as the mother in the two crosses. A classical example of difference between reciprocal hybrids is that of the horse-donkey hybrids. Here, the mule, the product of a male ass and mare, is supposed to differ from the hinny, the reciprocal hybrid. As Goldschmidt pointed out however, the cross in both cases is not one between the same two breeds. Mules are generally produced from heavy draught-mares, Percheronnes, or Mulassières or Belgian, and some very large donkey of the Poitou, Catalonian or other Spanish breed. The reverse cross however, is only made where donkeys are plentiful and cheap, and mostly very small, such as in Mexico or Algeria, and the stallion used is very often some small nondescript pony. If however, the same breeds are used in cross-breeding, that produce the valuable mules, hybrids are produced which are practically identical with the reciprocal hybrids. Hybrids from a heavy draught-horse stallion and good, large Spanish Jennies are hard to distinguish from mules bred in the ordinary way.

In the instance of the Oenothera however, the parents used for the reciprocal crosses, are the identical plants, and the hybrids from Biennis ovules and Muricata pollen are unmistakably different from those developed out of ovules of the same Muricata plant with pollen of the Biennis plant used in the reciprocal cross.

The only explanation is, that in one or in both species, the pollen and ovules are geno-typically distinct. Whatever the nature of the difference, it must be such as to persist through several generations. For both kinds of hybrids are generally stable for their characters, if self-fertilized. In further work with these hybrids, de Vries discovered a very startling fact, which up to a certain point gives an explanation of the peculiar mode of inheritance. Namely, that in the Biennis-Muricata hybrids, whose father was Muricata and whose mother was Biennis, all the pollen had the properties of the pollen of Muricata, and all the ovules had the properties of Biennis ovules. The correspondent identity of the pollen of the hybrids with the pollen of the father, and that of the ovules with the ovules of the mother was observed in the reciprocal hybrid. This came to light when the pollen of the hybrid was used in back-crosses with the parent species. The pollen of the Muricata-Biennis hybrids, whose father was Biennis, used in fertilizing Biennis flowers, gave normal Biennis plants, just as the real Biennis pollen would have done. The ovules of the same hybrid, fertilized with Muricata pollen gave wholly normal Muricata off-spring.

The result of this very curious phenomenon is that such hybrids breed true. In the Muricata-Biennis hybrids, the pollen produced is Biennis pollen and the ovules are Muricata ovules. Each self-fertilization therefore is a new hybridization, and again produces the same Muricata-Biennis hybrids. We have the following scheme:

Initial cross $A \times B$
Hybrids $A—B$
Gametes: A ovules, B pollen
Self-fertilization: $A—B$
Gametes A ovules, B pollen, and so on. One result of this state of things is, that when the reciprocal hybrids. $A—B$ and $B—A$ are crossed, one of the parent-forms results A or B, according to which way the cross is performed. This is certainly the most striking result, but the fundamental thing in these experiments lies in the demonstration that in these

11

species of Oenothera used, the pollen of a true-breeding form can differ geno-typically from the ovules. The breeding true of hybrids is demonstrated and partially explained, but at the same time it is shown that such apparently pure species as Oenothera Muricata and O. Biennis can be heterozygous for very many genes without having for that reason a hetero-geneous descendance. It is clear that we may expect any true-breeding Oenothera, even if it belongs to a recognized species, to be in this peculiar state. The fact that reciprocal hybrids between such species are far from identical, and at the same time breed true, shows that in every such an instance at least one of the species crossed must have pollen which differs geno-typically from the ovules. From the experiments of de Vries it is clear, that Oenothera biennis, Oenothera Lamarckiana, and Oenothera rubrinervis are all three in this condition. The pollen of a pant of Oenothera Lamarckiana need not contain the same genes as the ovules of the same plant. A true-breeding plant in this group may even be heterozygous for a certain gene if we regard it as a male, and heterozygous for other genes if re-garded as a female.

The interesting facts discovered by de Vries in his exten-sive work, show very conclusively, that here there is a peculiar mechanism, as yet unexplained, which makes that plants, heterozygous for several genes, may yet breed true.

And it is these facts, in my opinion, which furnish the key to the very peculiar phenomena de Vries had observed long before, in some of these species of Oenothera, notably in O. La-marckiana. The sudden production of new forms in this plant looks very much like mutation, and we know that de Vries has interpreted them as such, even to the point of generalizing this sudden production of novel forms into a mutation-theory of evolution.

Several authors, including myself, have tried in vain to in-terpret this "mutability" of Oenothera lamarckiana as a case of complex heterozygosis, with the resulting production of reces-sive forms, given in the genotype of the individuals. The dif-

ficulty in the way of this hypothesis has been the fact, that
such novelties were often produced only after the parent-
species had been grown for a number of generations without
producing any aberrant forms.

Now however, we know that plants of Oenothera Lamarckiana
may be complex hybrids, that they may be impure for quite a
number of genes, that their pollen may be geno-typically different from their ovules, but that they may nevertheless breed
true, as the result of a peculiar mechanism, which is as yet
obscure. This mechanism, this mysterious set of conditions,
produces the remarkable result that individuals may produce
pollen like the pollen-cell furnished to their make-up by their
father, and ovules like that from which they developed. What
happens in these plants must be some sort of a suppression
of the normal synthesis of the germ-cells. We might conceive of
some process analogous to that which takes place in the production of periclinal chimeras, in which two geno-typically
different individuals live in intimate contact, one inside the
other, and yet retain their geno-typical identity. We can imagine how a plant of Oenothera grows up as a composition of tissues of different identity, partly composed of cells directly produced from the original male gamete, and partly of cells derived directly from the original ovule. If then we have sufficient
imagination to think of a way in which this male individual,
derived from the pollen can furnish pollen, whereas the other
individual, derived from the ovule, in its turn will furnish the
ovules of what looks like one plant, we could make some working-hypothesis, which may prove of some use in the investigation of the peculiar mechanism. We would not be surprised if
some of these at first sight sexually produced hybrid Oenotheras proved to be a sort of periclinal chimera, formed in a quasi-
sexual but really asexual way, composite of the two parent-
forms, such as Muricata plants in Biennis epiderm, just as
Cyticus adami is Cyticus laburnum with an epiderm of Cyticus
purpureum.

One thing is certain, namely that the process by which Oeno-

thera hybrids are perpetuated if they are self-fertilized, which
process must be the same as self-fertilization in so-called pure
species, does not admit of segregation in heterozygotes. We
may say that in Oenothera the segregation which we see in all
other organisms if they are beterozygous, is suspended. We do
not know what causes underly this process, but we must remember that anything which disturbs it, must set free this
Mendelian segregation which was only suspended by it. We saw
that in several pigeons, where all the females are heterozygous
for a gene indispensable for pigmentation, and where no albinos
are formed because some unexplained cause prevents the formation of male-producing eggs without the gene, the spontaneous production of an albino in pure-bred coloured-stock
might not be called mutation. In the same way, now we know
that something suspends segregation in obviously heterozygous Oenotheras, we may not call by the name of mutation the
result of a segregation, which was only suspended for generations by a cause of which we know very little.

The assumption that Oenothera lamarckiana was of hybrid
origin, is in accord with its probable production as a hybrid in
the Paris botancal gardens. It does not grow in a wild state
anywhere, and for several years Botanists have in vain been
exploring the Southern States of America for it. The fact de
Vries finally sowed the seeds of Oenothera lamarckiana in the
United States in likely places, and thus made impossible further search for possible wild habitats, sufficiently shows how
even he has given up all hope of proving it to be an original
wild species.

Now we have to go one step further, and discuss the possibility of the production of a new, dominant form within a pure-
breeding strain of Oenothera, without the intervention of a
mutation.

It is very clear from all the work of de Vries and Shull and
Davis with reciprocal hybrids in Oenothera, that in some of
these plants there is some mechanism which links the ovule
furnished by the mother to the ovules produced by the daugh-

ter. And that there is independently from this, a mechanism which makes the pollen of the young plant identical with that of the father. Now, whatever the nature of this mechanism, it pre-supposes at least, that in some way the two processes do not interfere, just as the continuation of cell-division in the generations of Cytisus laburnum cells of a Cytisus adami tree does not interfere with the genotype of the generations of C. Purpureum cells in the inside. The characters of this graft-hybrid are the result of a super-position of a layer of cells of one species over tissues of the other species, but this something radically different from real coöperation of genes derived from two species, such as would take place in a real sexually produced hybrid between the two species. And in an analogous way, there may be a great difference between real hybrids in Oenothera and the sort of hybrids which show suspended segregation. But at the same time we can conceive of a result of the cessation of the process which is responsible for this suspended segregation, other than the production of recessive novelties, namely the production of dominant novelties, produced by combination of genes within the cells. The male series of cells in a true-breeding Oenothera may contain a gene A, and the female series of cells in the same plant, and the same succession of plants may contain B. But it is possible that these genes never meet in the same cells, so that they do not coöperate in the way in which genes coöperate which are present in one cell, and in many generations of cells. Just as recessive novelties must be produced at any interruption of the mysterious process which causes suspended segregation in Oenothera, so can we conceive of combinations of genes at the same time, combinations of genes which had been present in the same plant but not in the same cells.

It is very significant, we think, that the only instance of a production of a dominant novelty in plants outside crossing is the case recorded by Gates, the production of a red-fruited form in one of these same Oenotheras which show the remarkable true-breeding of hybrids. We would warn seriously against accepting such instances in this material as proving the spontaneous creation, or even the spontaneous acquisition of new genes, real positive mutation.

Other cases of dominant novelties in pure-bred stock are Morgan's cases in Drosophila. Drosophila, although it furnishes excellent material for genetic studies, has furnished all sorts of complications, irregularities in the ordinary segregation of genes, in some respects somewhat similar to the case of Oenothera. Then, we have to remember that dominance is relative, and that by loss of one gene, individuals may be produced which have a character, dominant to a corresponding character of others, which look like the original stock. In chickens, white animals may produce coloured off-spring, for instance, white Leghorns may produce brown ones. Such brown fowls have a colour, dominant to the white colour of the Silky, or of the white Rosecomb bantam. According to Punnett there are black rabbits that may produce agoutis, it they are heterozygous for a gene which they have more than the agoutis. Such agouti is dominant over the black of most black rabbits.

In other words, the production of such dominant novelties as pigmented animals in chickens or agouti rabbits, may be due to loss, or at least removal of an additional gene, which made their progenitors look like the recessives. In addition, the production of dominant novelties as the result of combination of genes, not heretofore present together, is possible in a more or less pure-bred strain. Such a combination may be noted only after several generations, if both individuals carrying A and individuals carrying B are relatively rare.

Only if we are absolutely certain of the purity of the individuals of a given family, we may without further tests call mutation the sudden production of unexpected novelties. If, however, we can be sure of the absolute homozygosity of a group of plants or animals, we can feel free of the burden of further tests.

In these cases where we have to demand the observance of the most rigorous precautions before we can accept the reality of an alleged mutation, we have an analogy to the controversy about "Generatio spontanea" in the time of Pasteur. Believers in this process urged that this spontaneous creation could not possibly have anything to do with the preparation of the test-tubes filled with food-media. It seemed obvious, that if a spontaneous growth of bacteria is possible on potatoe, it matters not at all whether this potatoe is raw or boiled. In the same way, we hear the statement nowadays, that if spontaneous geno-variation, the spontaneous loss of a gene is possible, it can happen as well in a family of plants or animals which are not strictly pure for their genes, as in a family in which all members are strictly homozygous. The difference comes in from the side of the necessity of control. Pasteur reasoned, that if bacteria were spontaneously created, they could be created on boiled as well as on raw potatoe, and the fact that spontaneous growth was never observed on well sterilized media was very significant.

Some authors still believe in the possibility of demonstrating spontaneous geno-variation in impure material, and without control-matings. One example is the so-called spontaneous production of a dwarf of Oenothera biennis by Stomps, a pupil of de Vries. This author crossed strains of O. biennis which differed in shape of the flowers. One family had extremely narrow petals, but otherwise the two were pheno-typically alike. In the second hybrid generation Stomps obtained two novelties, a giant and a dwarf, which he calls-mutations. Of course such an instance is an extreme case of abuse of the term. Very few authors nowadays would call the production of novelties in the second generation of a cross mutation, at least not in the sense of real spontaneous geno-variation, in de Vries' sense of the word. It is obvious, however that, if the two parents had not been different in flower shape, if they had been identical phenotypically, the case would have looked much more like a real mutation.

In fact, Stomps obtained the same dwarf forms from purebred Oenotheras, in what he calls a "pure line" of Oenothera. That is to say, wild-growing plants, inbred for four generations, produced this aberration. Now we know what breeding true is worth in this material as a criterion for geno-typic purity, but even in other material, the sudden production of a novelty after four generations, comprising only very few plants in every generation could hardly be called mutation without breeding tests.

We know, that in several instances the difference between individuals having alternative characters is due to the presence or absence of not one definite gene, but of at least one of a set of two or more genes which have a similar action on the development. In our comparison of the chain of processes leading to the development of any quality of an organism to a material chain of metal links, we can imagine how in some place two links are not held together by one single link, but by two or three or more links, all passing through the same upper link and through the same lower one, lying side by side. Ordinarily, when one link breaks, the chain breaks at that point. If in any place on the chain however a duplicate link is put in, a link

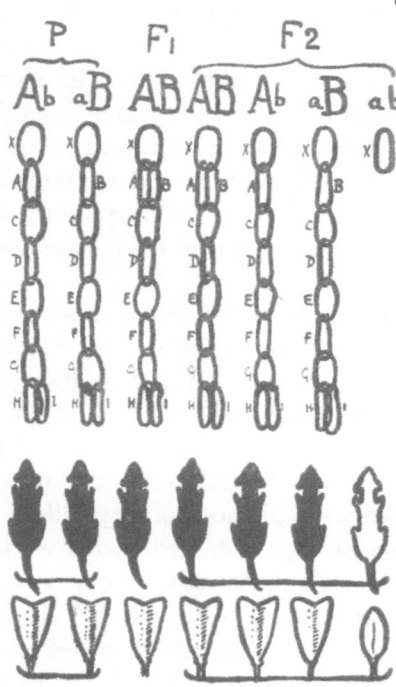

Fig. 20.

The effect of "duplicate" genes, genes which affect the same stage in a chain of processes in a similar way, and are therefore, in respect to the end-result, of equal value.

Only when both A and B are missing the following links are unsupported.

can be taken out without interrupting the chain. (Fig. 20). The chain remains intact until at this point both links are removed, or all three as the case may be. The very best example of such an instance is the one which Shull discovered in Capsella. Shull found that several normal Capsellas, with triangular fruit, differed from a form with roundish fruit (Heegeri) because of the possesion of three distinct genes more. He further found, that each one of these three genes was in itself sufficient to produce the difference, sufficient therefore in this instance to "determine" the triangular shape of the fruit. If we call these genes A, B and C, plants which have only A grow up normal, and so do plants with only B or C, or any combination of these three genes. In fact, only those plants which lack all the three genes have the roundish fruit. The result is, that if Shull crossed a plant with normal fruit to one with the oval capsules, the hybrids had normal fruit, and the proportion of "Heegeri" plants segregating out of the second generation varied according to the presence in the normal grandparent of one, two, or three of the genes A, B, and C. Some F_2 families would give one plant with oval fruit out of every four, others produced one plant Heegeri out of every sixteen, and in other families only one individual in every sixty-four would reproduce the recessive character. Honing found an analogous case in Canna.

Now one of the results of this fact, that either of three different genes is sufficient for the production of a certain dominant character, is, that the recessive character will inevitably be produced in F2 if two plants are bred together which are both normal, but do not happen to have either of the three genes, A, B, or C in common, for instance if we cross a plant with A to one having B and C only. And we will see how this cause of the production of recessive novelties from parents, pure for a corresponding dominant character can be very easily mistaken for real spontaneous loss of a gene, real loss-mutation. In fact, it is only prudent to say that in all animals and obligatory allogamous plants only very elaborate test-matings can make an alleged case of mutation stand as proved.

In Shull's work with Capsella, the number of parallel genes

all having the same effect upon a certain stage of the development of the young fruit, was three. But it is easy to conceive of cases where four or six or more genes, which in their action upon the rest of the development may act quite differently, have the same influence upon one certain point. Any day Shull may hit upon Capsella plants lacking A, B as well as C, and still with normal capsules, because of a possession of a gene D, or two genes, D and E.

Now in all such instances, wild-growing heterozygotes may be taken up into our experimental gardens, which when self-fertilized will produce a very small proportion of plants with a recessive new character. If the number of parallel genes is large, and the number of plants grown in each generation is small, it may take two or several generations for the novelty to be seen. In a case of five genes, only one plant in every 1024 descendants of a heterozygote would show the new recessive character.

The required test-matings which, as we have just seen, are very difficult to perform, and which it is sometimes impossible to make, as in annual plants, can nevertheless hardly ever be dispensed with if we want to make sure of a case of mutation. Mutations in pure lines however, can be accepted as such. The difficulty here lies in the fact that it is hardly possible to be sure, that a group of organisms constitutes a pure line. If we hold to Johannsen's definition, a pure line is a group of plants grown by repeated self-fertilization out of one individual, pure for all its genes. The rapid reduction of geno-variability caused by self-fertilization makes it probable, that nearly always the descendants of one individual plant of wheat or oats or pea is a real pure line. And, therefore, spontaneous production of novelties in such material, as observed by Johannsen and by Nilssen Ehle have great significance, although it is not strictly defendable even in this material to speak of pure lines. And we should, to avoid confusion, always protest against the abuse of the term in those cases where authors denote by "pure line' simply pure-bred material, such as self-fertilized series of plants, or worse, vegetatively propagated material, or Oenotheras, the geno-typic purity of which is not investigated. An extreme case of the misapplication of this term is certainly Blaringhem's use of the word, speaking of sheep and guinea-pigs!

There can be only one exception to the rule that test-matings are necessary to show the homozygosity of the material, to conclude that the sudden production of a novelty is a real mutation, a real spontaneous change of genotype, acquisition or loss of a gene. And this exception is the case in which we know that a certain individual must, because of the very mode of its origin, be pure for any gene it contains. If an animal or plant develops from one single gamete we think it could not possibly be otherwise than homozygous.

Really parthenogenetic individuals must be pure, and they should furnish the only irreproachable material for experiments with pure lines. Such animals as in the experiments of Loeb developed out of real fertilizable eggs of sea-urchins and frogs, cannot be otherwise than homozygous.

In 1912 we started a set of experiments with Squashes and Marrows. In these plants it is possible to obtain fruit containing good seed from unfertilized female flowers. Our method was to close the female buds with lead-wire, removing the male buds on all the plants every evening. We performed several hundred cross-fertilizations between widely different forms. The hybrids, grown in 1913 proved to be in every instance different from the mother-form, thus proving the absence of apogamy. Reciprocal hybrids were in every instance identical. On these hybrid plants we closed the female buds. In doing so, we found that several hybrids would never produce any fruit from closed buds, though they would later in the season produce numerous fruit from fertilized flowers. Other plants produced parthenocarpic fruits, apparently normal, but void of seeds. Still others yielded good fruit with empty seed, and finally, there were some hybrids which produced normal fruit, full of viable seeds. It is to be noted, that in no instance did we find a fruit with only a few good seeds, such as would probably.

have resulted from a faulty technique, and the presence of pollen on the stigmas. The fact that only certain combinations
would produce good seed from unfertilized female flowers considerably narrowed the range of our crosses. In 1914 the seed
of three non-fertilized hybrids was sown. We expected either
of two things, apogamy, production of seeds by some asexual
process, from cells not identical with real fertilizable ovules
or real parthenogenesis, the spontaneous development of real,
fertilizable ovules into seeds. In the first case we would have
obtained a descendant which reproduced the mother-type, the
characters of the hybrids, and in the case of real parthenogenesis wee xpected to see a Mendelian segregation, the segregation
of genes over the gametes of a heterozygote. The latter alternative possibility proved to be the case. The cross between
Miracle and Vegetable-marrow may be cited as one of the clearest instances. Vegetable-marrow has long, narrow fruits, with
perceptible ribs. The fruit has a hard shell which is white, and
turns bright yellow on ripening. The seeds are normal. Miracle
has fruits which are very much heavier, roundish-oval in shape,
flattened at both poles. There are no ribs, and the colour is
dark green, marbled with yellow stripes. The fruit has a very
soft shell, which even in ripe fruit can be pierced with the
finger, and the seeds lack the seed-coat, the cotyledons being
naked.

The hybrids had oblong fruit, shorter than Vegetable-marrow, but not as round as Miracle. The colour was green marbled, with yellow stripes. Ribs were absent. The skin was hard,
and the seeds were all covered with normal hard seed-coats.
Both reciprocal hybrids presented the same combination of
parental charcters.

Among the plants raised from unfertilized ovules, only twelve
were grown until they produced ripe fruit. Of these, seven
bore green fruit, more or less marbled, four whitish yellow, and
one orange with yellow stripes. Five of them showed ribs, and
seven had a smooth top. Four of the plants gave fruit with
naked seeds, and the other eight had seeds with seed-coats.

Most of the plants bore oval fruit, and there was no plant which reproduced the exact shape of either grand-parent. Two of the plants had decidedly pear-shaped fruit, two had almost globular fruit, and the plant with the orange fuit happened to have very long fruits, which may have differed in shape from Vegetable-marrow, mainly because they were soft-shelled. In size there was a great variation in the fruit of the different plants, but as the plants were grown climbing on stands, whereas the original plants and the hybrids were grown straggling on the ground, no fair comparison can be made on this point.

This segregation shows clearly, that, whatever may be the nature of Mendelian segregation at the formation of germ-cells, these seeds must have passed through the process. In other words, true, fertilizable ovules must in this case have developed spontaneously into viable individuals, good seeds. As each of these individuals must have come out of one gamete instead of two, each of them must have been pure for all its genes, and any group of plants grown from such an individual must be a real "pure line." In such material further tests would be superfluous, and any spontaneous hereditable variation in a self-fertilized series with this origin might without further investigation be termed a mutation.

We found the climate of Buitenzorg, Java, deadly for our squashes. All our seed was dead within fourteen months, and the two series of young plants from self-fertilized seeds of parthenogenetic plants could only with very great care be protected long enough against fungi and insects to show the shape of the young fruit, and of the leaves. Thanks to the care of Mr. van Helten of the Botanical gardens, we could observe that in the descendants of a parthenogenetic F2 plant from the cross Türkenbund x Poire bicolore there was no variation in fruit-shape and during our illness, Dr. Smith of the Botanical Museum grew a number of plants from self-fertilized seeds of one parthenogenetic F2 plant Miracle x Vegetable-marrow for us, and reported uniformity in fruit-shape and leaf-shape.

To resume our chapter on mutation, we have to conclude

that it is very probable that mutation, at least loss-mutation is a phenomenon which occasionally occurs. We have tried to show, what extreme difficulties lie in the path of the genetician, who wants to obtain certainty, that the sudden production of a novelty is not caused by redistribution of genes over the gametes of a heterozygote individual. We saw that such a redistribution of genes, Mendelian segregation, can in some instances be deferred for several generations, making test-matings necessary in almost every instance.

If we exclude as doubtful all those instances of alleged mutation in which the necessary tests could not be, or simply were not made, there remain only five or six instances. And as those instances were all cases of loss-mutations, or at least cases in which a recessive novelty was produced, which we have tried to show means loss of a gene, we can only conclude that the rôle mutation can have played and can still play in the evolution of species is at the most a very insignificant one.

SELECTION.

ELSEWHERE we have discussed the different factors in evolution, and the circumstance, that on the whole those factors can be grouped in two categories, factors which heighten the potential variability of groups, and such as reduce the potential variability.

In regard to species-formation, we can say that those causes which heighten the Total potential variability of a group, produce the conditions required for the possibility of the origin of new species; and that those causes which reduce the Potential variability, will make a species of any group of organisms under certain conditions of relative isolation.

How does selection stand in this respect, to which category must we bring it? Is selection always a cause or reduction of the Total potential variability, and thus a factor of some importance in the establishment of species?

Darwin, and later Weismann tended to see the cause of specific purity and the reason for its continuance in natural selection. Natural selection was thought to keep a group of organisms with a natural tendency to vary in all directions down to a limited variability, by a weeding-out process, which would tend to conserve only the individuals which were best adapted to the conditions under which the species lived.

We have tried to show, that a group of organisms which is limited in certain ways, which is so situated that the proportion of matings with individuals from outside remains below a certain maximum, as compared to matings between its members, will automatically reduce its potential variability. It is a matter of relation between the influence of factors heightening the variability (crossing) and factors reducing it, (isolation,

autogamy etc.) whether the total variability of a group will
continue to reduce itself, whether it will remain approximately
stationary, or whether the status of the group as a species
is insecure.

Theoretically, the automatic reduction of the potential variability of a group goes on, independently of any selection.
But the advocates of natural selection as the main cause of
evolution, have shown that, as only a fraction of the number
of individuals produced in any group, have a chance to grow
up, it is natural to suppose that the survivers are a selected
group, and are in the main surviving because they are better
fitted to their environment. We will later discuss whether the
life or death of the individuals really depends, on the average
of their constitution and on their suitable situation.

In the first place, is it absolutely necessary to assume that
very organism which we observe to be well-fitted to the circumstances under which it exists, has been made fit for those
circumstances by natural selection? In most cases there seems
to be no mysterious force compelling a group of organisms to
remain in an environment to which they are only moderately
well-adapted, when another environment is open to them into
which their present constitution makes them fit better. Either
an organism happens to have a genotype which makes it so
constituted, that it can live and procreate in a certain set of
conditions, in which case that particular organism will be
found to be living right there, or else it cannot exist at all.
Adaptation, is certainly not the right name for every case in
which we find an organism living in surroundings in which its
germinal constitution allows it to live. On the other hand, it is
perfectly obvious that there are cases in which a plant or animal is so constitued, that it is securely tied to special conditions, such as a fish to life in water.

On the whole, the striking suitability of organisms for the
conditions in which we find them, has resulted far more from a
selection (by the method of trial and error) by the organism
of a suitable environment, than from the selection of the organ-

isms (by the method of selective elimination) by their environment.

The ultimate future of a given group of organisms, the possibility which it has of becoming a species, depends upon the equilibrium betwen the two sets of causes influencing its potential variability. If the situation of the group and its constitution are such that the balance is in favour of a greater variability, it will not be a species. If on the other hand the factors, which reduce its potential variability, are more potent than the causes for a heightening of the variability of the group this group should be considered a species for that reason. What the contributing factors are, whose cumulating action produces one effect, or the opposite, is to a certain extent immaterial for the final result. Unless for instance, slow reproduction, which makes for greater potential variability than rapid reproduction, is off-set by geographic isolation or any other effective isolation, the formation of numerous species in one territory is hindered by it. A great tendency to promiscuous mating may be off-set by rigorous artifical selection. We are concerned in the first place with the balance of tendencies toward and against reduction of the potential variability.

It depends greatly on the nature of the material, and the nature of the selection, whether this last factor will materially affect the variability of a group of organisms. In some cases selection acts as a means of isolation, and tends to reduce the potential variability of a group in the way in which all sorts of isolation must act. The most striking instances of this are seen in the domestic animals and plants, where in a good many instances the matings are promiscuous with the exception of groups of selected individuals, which are so protected from intermixing with organisms of other groups, that they constitute species. Only relatively few of the individuals of such species have off-spring which can be counted into the species, only individuals with certain merits, individuals which live up to the fancier's pre-conceived ideal of the species are chosen to reproduce it. Many others will eventually produce off-spring, but as their breeding is uncontrolled, their get falls outside the limits of the species. In respect to the effect of selection these domestic species differ somewhat from species in a state of nature, and it is necessary to consider the nature of the difference as well as the points where domestic and wild species are affected by selection in the same way. This consideration will be illuminating chiefly in the question, how far we can generalize the phenomena observed in the evolution of domestic species.

It is possible to develop a new breed of dogs and to bring it to perfection in a country over-run with mongrel dogs of all kinds, simply by choosing suitable males for certain females and controlling their mating. Our choice of individuals can be made according to any preconceived idea of what we want to preserve in our dogs, what we want to combine, what should be excluded. By controlling our animals at mating-time, the fitness or survival-value of the characters selected by us need not be taken into consideration at all. We are, moreover, not concerned with the constitution of the dog-population at large, in the region where we are perfecting our species of domestic dogs. No group of animals or cross-fertilized plants can ever grow to the status of a species, unless it is either so constituted or so situated, that matings within the group are far more numerous than outcrosses.

In the case of the domestic species, selection is a means of strict isolation in addition to its influence on the constitution of the selected group. Natural selection must necessarily act qute differently in different situations, even if we can conceive how in the long run, and speaking statistically, it will tend on the average to act by favouring a certain type. But artificial selection is selection according to one ideal, and it is acting as a directing factor just as strongly and efficiently in favour of an altogether irrelevant quality, or even of a decidedly harmful character, as in favour of a quality which makes for the success of the individual showing it.

If we compare two groups, which are each sufficiently effectively isolated from admixture with individuals which do not belong to them, one can be isolated as the result of an actual geographic separation from individuals which could cross with its members, or as the result of some geno-typic quality which makes the individuals succeed only in situations where individuals not of its species are not or seldom present. Another group may be isolated wholly by the wishes of breeders, who control the breeding of the group to make its members conform to a certain standard. Both these groups are species, the first a natural species and the second a domestic one. Both have this in common, that the situation and constitution of the group is such, that the potential variability cannot but decrease.

Either the potential variability of such a group is zero or as near zero as is consistent with the differentiation into individuals of different sex, or the potential variability of such a group attains a certain magnitude. If the latter is true, further reduction of the Total potential variability of the group is still possible, and as a result of this fact, the eventual type, the eventual geno-typic constitution of the group is not yet rigidly set. It is conceivable, that in such a case there are no phenotypic alternatives given in the constitution of the groups, but it is clear, that very often the eventual type for which a group can become pure is not as yet rigidly laid down. In such groups, selection may affect the outcome, just as much where the reason for the inclusion of certain individuals among the number which produce off-spring rather than others, lays in their greater chance to develop up to reproducing age (natural selection), or in their greater comformity to the breeder's ideal (artificial selection).

But we must always bear in mind, what we considered in the chapter on the reduction of variability, namely, that minorities have no chance. Natural selection cannot do other than affect the average chance of survival of certain individuals. We can take a simple illustration, a certain variation in size in a group of organisms, which partly depends upon the geno-typic constitution. We can calculate the chance which any individual of this group has, to be amongst those who are parents of the next generation. Other things being equal, this chance depends upon the rate of reproduction of the group, upon the stability of conditions favouring development of organisms of this kind, and greatly also upon chance combinations of circumstances around every young individual. If for some reason, the large individuals have some additional advantage over the smaller ones, the chance for a large individual to be amongst the parents of the next generation will be accordingly greater. If originally both large and small individuals happened to exist in approximately equal proportions, it is evident that the better type, the larger type would tend to become the specific type, the type of the species, if the variability of the whole group tended to reduce itself for any reason or combination of reasons.

But if the large individuals are far in the minority, this low proportion of their number to the total number has to be reckoned with. The chance for a large individual to be one of the pro-creating individuals depends for a good deal upon the proportion of large ones to small ones, and if this chance is materially increased by a greater fitness, we are still concerned with the modified, low proportion. If now we consider artificial selection, we must see that if a breeder considers great size of some advantage, he will isolate a group of large individuals. Some of these may be impure, of course, and as the choice may happen to be wholly on phenotypic merit, many individuals which are potentially small ones may be included, but it is evident that by this isolation of a number of large individuals, the chance of the eventual type of the selected group to be large is very greatly enhanced, out of all proportion of what can happen along this line by natural selection without isolation.

And this consideration of the fact that selection under domestication is effective through the fact of being isolation as much as selection shows us, where and when we have to look for the action of natural selection in the moulding of the type of species.

To a student of evolution it is apparent how domestic species, dogs, poultry, sugar-beets, change under the action of selection, and at first sight it seems logical to assume that selection must gradually modify wild species in approximately the same way. The essential thing, however, which distinguishes domestic species from wild ones, is the existence in wild species of great blocks of individuals of one type, whereas in domestic species, the group of individuals chosen to be the parents of the next generation is always a very limited one, of which each member is carefully chosen according to one standard. These few selected individuals are not left as a minority in a mixed population, but they are themselves isolated in a group in which they constitute the type, the ma ority.

When a relatively small group of individuals of one species splits off, such as is constantly happening, where seeds and young animals wander into surroundings where somewhat earlier the species had no representatives, the future of this group as a possible new species depends upon several things. One of these is the genotype of the inidividuals; if the group is really representative of the old species, and consists wholly, or almost wholly, of normal, average individuals, it will never be able to develop a type of its own. Another important factor is isolation. Given a sufficient potential variation, the new group in getting pure, may become so for a somewhat different set of genes than that of the average individual of the old species, but this can only happen when for some reason, inter-crossing of its individuals with members of the old species is relatively rare. Geographic isolation is of an obvious kind, but isolation can result from the very constitution of the group, as well as from its situation. Any group, sufficiently isolated, must lose its total potential variability, which means that it must eventually become pure for its own type. And it is apparent, that at this very point natural selection must make felt its influence. Elsewhere we have calculated how great the chances are for the loss of some types, which occur together in one group. Minorities tend to disappear in this way. If we suppose, for the sake of simplicity, that the potentialities of a small group are such, that either one of two types can eventually become predominant and exclusive in the group, what are the factors which actually determine the outcome?

Other things being equal, the numerical pre-pronderance of one type, or to be more exact of the gametes produced having a certain gene over the number of gametes produced without it, will decide the eventual type. At any moment at which only a few individuals happen to be the parents of the next generation, the chance that the production of one kind of gamete is within the potentialities of these few individuals is greatest for the majority type. But on the other hand, numerical proportions being equal, greater or less fitness of constitution for the given environment will decide. Natural selection will in those circumstances become an important factor.

In those special circumstances, when relatively small groups are living out of touch with the main body of individuals of closely-related species, natural selection must have an importance in deciding the outcome of the natural experiment in differentiation, which it cannot have elsewhere.

We have common species and rare species, and we have species which are common in certain localities and rare in others. This must mean, that their constitution makes it possible for them to live only, where we observe them.

But we know, that there is a continual striving of seeds and young animals to try life in places, which are not suitable for their constitution. Conditions change, and often we can notice how certain localities are apparently very suitable for certain plants or certain animals without being occupied by those organisms. Eventually, chance will bring them there, and it is clear that the less likely this is to happen, the better will be the temporary isolation afforded to the small group. The chances against the evolution of new plants and animals, at least for as far as we are concerned with allogamous plants and animals, are heavy. Individuals with a new genotype, which result from a temporary heightening of the potential variability of the species by a cross, have no chance to change the type of the species, or to found a new species in the midst of the old one.

New, isolated colonies of plants or animals will simply be colonies of members of existing species, unless the isolated group should happen to have a markedly high potential variability. For the formation of a new species, no matter of how ephemeral a standing, a combination of two circumstances is necessary, colonization, or some other cause making for isolation, and a typical variability of the isolated group.

Cases in which we can notice this combination of two relatively rare circumstances, have been observed not to infrequently. They can be said to be instances of the origin of species. In most cases the causes for isolation are not of a permanent nature, and there the newly formed species has only an existence of very short duration. Two instances will illustrate what I mean.

Colonies of the Norway rat are being formed almost inevitably where conditions of shelter and food-supply happen to be favourable. These colonies consist almost always of normal, typical Norway rats. But at least one instance has become known of a colony of aberrant rats, black instead of grey, which show the white markings on feet and belly of Mus norvegicus much more plainly than the typical rats. This colony was observed in Ireland in Co. Wexford by Rev. R. Keating. It was a species of very temporary existence as a wild species, but as a domestic one it is still being bred in cages by rat-fanciers under the name of the "Irish" rat.

Another similar case is the occurence of a species of very marked black rats, Mus rattus, which were found to exist on board a steamer going from Buenos Aires to Amsterdam. It is probable that this species developed as such, on board the steamer, possibly out of some hybrids, or from rats which came on board at different ports. These rats were very small,

somewhat intermediate in size between Mus rattus and Mus
concolor, black in colour, with a very pronounced brilliant green
lustre, and they had extraordinary long tails.

Lloyd in his books on the "Growth of groups" gives numerous i nstances of such small, uniform, aberrant populations
of rats, which can only be classed as decided species of a very
uncertain permanency.

In all these cases, the species existence will come to an end,
when the special causes which afforded it the necessary isolation from inter-crossing with the multitude of typical individuals, will cease to exist. And this circumstance shows us
that in the origin of species mere temporary isolation of a
small group with a high potential variability is not enough.
The combination of causes, the chance, which brought the
first colonists to the spot where they could found a new tentative species, will eventually bring so many individuals of the
old species there, that the new group will cease to exist. And
it is clear that a species can only keep on existing if it can extend its range, so that adverse circumstances, which annihilate
it in certain regions, will leave sufficient numbers of individuals
to continue the species. A new species which differs from a multitude of typical individuals cannot extend its range into the
territory of related forms, unless there are circumstances
which hinder free crossing.

From the work of the systematicians with such animals as
small rodents, it follows that such local species exist, which
cannot coëxist in one locality, which differ from each other in
non-essential, non-adaptative characters, and which are
bound to a certain more or less circumscribed area by the
existence of other local species of the same group in neighbouring localities. In these circumstances it depends simply upon
the number of genes concerned in the differences envisaged,
whether we will find sharp or gradual demarkation between
the characters which distinguish these sub-species.

On the other hand, it is possible, that a group of organisms
which is temporarily cut off from random crossing with the

multitude, will become pure for a genotype which will afford the individuals having it a possibility to live in somewhat different conditions as compared with the related species. In other words, it is possible for a small group of organisms with a high potential variability, to come to fit into a different ecological niche. *This, evidently, is what we should term natural selection.*

If a few seeds from hybrids between Lychnis diurna and L. vespertina happen to reach a wet, shady, swampy, spot, it may be thought possible, that one or two plants of this lot would have a genotypic constitution which made life in these surroundings possible. We know how greatly variable in every point such progeny of species hybrids is. In this way under the influence of natural selection, a new Lychnis species could be thought to originate, for in those surroundings, the group would be so effectively cut-off from inter-breeding with either the multitude of Lychnis vespertina individuals, or with the multitude of Lychnis diurna plants, that it would have a chance to work out its own destiny as a distinct species. This view of the action of natural selection at the time of the origin of new species is wholly consistent with the facts pointed out by Wagner and later again by Jordan, namely that very closely related species are found, either separated by geographic barriers, or in distinct "ecological niches", and that only those groups of species cöexist in one "ecological niche," which do not inter-breed.

To cite a few examples: The field-rat of Java has been proved in our breeding-experiments to be at least geno-typically different from the Sumatra field-rat. All over Java there is only one field-rat, only one tree-rat, and only one big house-rat, all closely related, and occasionally inter-breeding, but fitting different ecological niches.

The small house-rat and the big house-rat, Mus concolor and grise venter do occur together in the same houses, and they live identically the same life. These rats, though obviously closely related, remain separate species because of their different size, which prevents cross-breeding, if not absolutely, at least' to an extent which would be inconsistent with specific identity. In the fields the field-rat exists together with a small rat, which completely copies its habits and mode of life, and which therefore is probably not identical with the Mus concolor which occurs in the houses.

When a species becomes stable, and pure for its type, time alone will decide whether it will be able to persist. It will only persist as a species, if it is somehow protected from merging into numerically larger groups, either by its geographic isolation, or by any other kind of isolation, a structure of the sexual organs producing self-fertilization, a great difference in size, in time of flowering, or a constitution which makes its members live a life which does not bring them into close contact with their relatives. And on the other hand, it will only survive if its individuals as such are well fitted to increase their kind.

In almost every natural order of plants and animals, new, small groups, differing geno-typically from the accepted constitution of the established species must continually get into conditions, where they are for the time being potential species.

There may exist species of animals or plants, relics of former generations which are irrevocably shut-off from any chance - of a heightening of the potential variability, by the fact that all their individuals have the same genotype, and that no species happen to exist with which they produce fertile hybrids. Ginkgo biloba may be in this condition. Generally speaking, however, the chances for the production of groups, which have such a potential variability, that they can eventually become pure for a new genotype is open in every group of species or sub-species, local species, within which cross-breeding is not excluded.

And as colonization is a normal course of reproduction in the most diverse groups, in some habitually so, in others at least occasionally, this trying out of numerically small groups which are potentially new species, must be an altogether common occurrence.

What we call natural selection is after all nothing but this process, the final outcome of such experiments in species-formation. In almost every instance the experiment fails. A small colony of individuals may have a potential variability which includes the possibility of a new genotype with a corresponding new character (or without), but eventually purity may be reached for the geno-combination of the parent-species, in which case this colonization differs in nothing from ordinary colonization. In other instances the group actually attains to a distinct genotype, but the causes, which brought about the original colonization, later on bring large numbers of the parent-species into contact with the group, which is accordingly swamped.

It is also easily seen, how circumstances during the first few generations favour a certain type, which accordingly becomes the type of the new species, whereas this group, which now has lost its plasticity, cannot continue to exist throughout the range of variation in the conditions of life through many generations. So that finally we see, that the chances for the succesful establishment of a new species narrow down considerably. For a new species to become established, a series of conditions must be fulfilled.

In the first place a group can only give rise to new species, if the range of genotypic possibilities is not exhausted in the material on hand; groups of animals and plants have been brought into cultivation for instance, which possessed no Potential variability, and from which therefore no new domestic species can be derived, such as the guinea-fowl and the Reeve's pheasant. Cross-breeding between geno-typically different groups is a first essential for evolution of species. For the production of varieties this requirement suffices, heightening of the Potential variability of any group, any species, necessarily leads to the production of varieties. But for the production of species, and for the change of varieties into species more is needed.

In the second place, only isolation of some kind can produce species, after the first condition has been fulfilled. For selection alone or fitness alone — cannot make a species out of the most happily constituted variety without isolation. In domestic plants and animals we saw that selection meant isolation, and it must be remembered that in nature, selection, that is survival because of merit, may become a means of isolation, and so a factor in species formation. In some instances special ability to survive in conditions into which the bulk of the individuals of the species cannot penetrate, can produce an isolation in space. This sort of isolation is obviously much more effective than chance isolation in space without special constitutional adaptation for this new habitat, for in the second instance swamping by the multitude of the old species is only a matter of time and chance This, to our way of thinking, is, where natural selection tends to influence the constitution of the surviving species. Only a constitutional difference from parent-forms sufficient to produce a preponderance of matings between members of the group over matings with individuals of the old group can isolate a new species sufficiently to make it survive in close proximity of the old one.

Autogamy is the most common mode of complete isolation thinkable. Where self-fertilization is the rule, and the same is true for any other process of multiplication without amphimixis, such as parthenogenesis and apogamy, new groups of organisms which to all practical purposes are species, will be found to abound. Habitual self-fertilization, or habitual apogamy will tend to restrict the crossing, and will therefore counteract a heightening of the potential variability. But in these groups, every individual, no matter what his genotypic constitution, is a potential species, of the same rank as a group of allogamous organisms which has under the influence of some kind of isolation reduced its potential variability and has attained a type of its own.

Selection in the case of these autogamous organisms means survival or dying out of these small species. Those which are best adapted to the conditions into which a mixture of several of them happen to live for a number of generations, will tend to survive, provided their numerical proportion to the total number is sufficiently large to counteract the tendency of a minority to disappear automatically. In these organisms new species, well adapted to life and to reproduction of their kind have a far lesser tendency to disappear; there is no multitude of individuals of one common type to mate with any surviving individual of the new type, and to compete with mates of its own type. The result is, that whereas the reduction of variability within each species, within the descendants of one individual in these autogamous organisms is extremely rapid, as compared to the purification of the type in an allogamous species, we actually find that the groups of individuals, which are generally reckoned as species in these autogamous organisms, are very polymorph compared to species of allogamous organisms. The explanation of the apparent paradox is simply that such species of plants as Triticum vulgare, Draba verna, are compound species, aggregations of numerous species each practically without potential variability, in a mixture. These are the "petites espèces" of Jordan, the "pure lines" of Johannsen.

The discovery that in a good many cases of polymorphy in plants, apparent species in reality consist of a great many small species of a high degree of purity, was one of great importance for an insight in evolution. But it is remarkable to observe here once more, the tendency of certain naturalists to over-estimate the importance of a new discovery.

The director of the Svalöf seed-firm, Nilsson, and Prof. de Vries postulated the origin by spontaneous variation, mutation, of these pure lines in wheat, barley and oats, and in the existence of so many pure strains in these cereals they saw proof for the idea, that in "mutation-periods" great numbers of new species originate spontaneously from one common parent-species.

Other authors have generalized the "pure-line-conception"

and want to see in "pure lines" the ultimate constituents of all species.

Blaringhem, in his lectures at the Sorbonne went so far as to speak of "pure lines" of guinea-pigs and of sheep.

Lotsy has attempted to identify species in general with pure lines, and, parting from the observation that pure lines have no genetic variability, defines species as groups of organisms which are devoid of genetic variability.

Contrasted to the marked polymorphy of such compound groups of very pure species, and of compound groups of vegetatively reproducing lines, clones, the relative conformity to type of the multitude in allogamous species is very striking.

Variability in such species of allogamous organisms may be relatively high, and yet we find on analysis that there does exist a common type to which the multitude conforms, and from which aberrant individuals depart in one or in a few characteristics, but seldom in very many. Within such species random mating counteracts the continuity of typical groups.

There is a very great difference between the kind of species which we observe in the autogamous plants, small groups, devoid of any Potential variability, and the sometimes highly variable species in the allogamous organisms. And yet, they are both the smallest permanent units. It is absolutely necessary to recognize in plants as well as in animals common, comparable units. If Systematics had to do without the species-conception, or if under the term species were understood different things in different groups of organisms, systematics would return to the pre-Linnean chaos.

For this reason any attempt, such as Lotsy's, to restrict the use of the term species to some special kind of species, which has no equivalent in other divisions of the organic world, should be discouraged.

In certain plants there exist small species, which are wholly devoid of variability of a genotypic nature. But this purity is not the essential thing, which makes these groups species. It is the result of the very severe isolation, produced by the structure of the flowers which ensures auto-fecundation.

It is right to call these groups species. But we must recognize that, whereas the purity of these species is very striking, it does not determine their status as species. While new lines of such plants are differentiating after a cross, they have not yet attained to purity, but they are already species. Should we restrict the term species to pure lines, or to some other special category of species, it would become necessary to find another terminology for all other species.

It is evident, that for as long as it was thought that species and all other groups of organisms had a natural tendency to vary, the importance of selection for species-formation was thought to be much greater, than we now know it to be. For we know now, tht there is no tendency to vary innate in groups of organisms. Variation is never spontaneous.

Even in the quasi-spontaneous cases of mutation, we have good reason to assume a cause outside the organism. Variation is caused by crossing. Closed groups lose their variability automatically, and we need not invoke natural selection to account for the eventual purity of the species.

In practical plant and animal-breeding our modern biomechanic conception of heredity as the transmission of genes, which may be factors in the development of both parents and off-spring, has not contributed much of a positive nature.

The best we can say for it is, that we have been able to prove to our satisfaction why certain methods of selection have been more effective than others. In most instances the breeders have empirically found the right way, and it is only rarely, that we Geneticians can find a case, in which we are able to point the way to the breeders.

Practically everywhere, the plant-breeders have abandoned their selection according to individual merit, which is selection according to phenotype, and resorted to some system of judging plants after their progeny. The whole elaborate system of selection of sugar-beets amounts to a comparison of the progeny of different plants, with a system of checking designed to eliminate as far as possible the influence of non-genetic developmental factors upon the choice of the best family. The system of selection of wheat and oats, which consists of a comparison of the progeny of a great number of individual plants and which was originated by Louis de Vilmorin is more than half-a-century old.

The animal breeders have been more slow to see the importance of judging an individual according to his get, rather than according to his individual merits, although the breeders of horses have always recognized the merits of individual sires as stock-getters, and patronized them accordingly. In comparison with plants, these domestic animals represent such a much greater value, and they have so much more importance as individuals, that this cannot surprise us.

It seems so obvious, that a bull of a breed of milk cattle is kept solely as a producer of good daughters, that the idea of Solomon Hoxey of inscribing bulls in a special Register of Merit, if they have produced a certain number of daughters of outstanding merit, would be thought to appeal at once to the breeders. And yet, we find important breeder's associations which judge bulls wholly according to their external characters — according to individual type.

Often enough, one can meet the old conception of heredity as the transmission of characters from parent to off-spring, in warnings against the use of breeding animals at a time when their indidivual characters are not at their best. Rabbit-breeders firmly believe that the use of a male for breeding at the time when he is moulting, is fatal to the quality of the coat of his otf-spring. We have witnessed an instance, in which the official permit necessary for use as a stud was refused to a stallion, because the animal was lame, whereas the committee who did the judging, had witnessed the accident which resulted in the lameness.

It is evident that there is an inducement to judge an animal according to his individual merit, namely that this procedure saves trouble and time.

In the practice of poultry-breeding, several systems have been proposed for judging an animal's value as a breeder by mere inspection. Some of these systems are very elaborate. For one of these systems it is claimed by the inventor, that by adding and multiplying certain proportions of a male bird, it is possible to find out how many eggs the animal would have laid in a year if it had been a hen!

The usual system of selection for egg-production in poultry is based on the records of females exclusively. Hens are valued as breeders according to the number of eggs they lay, and male birds are selected with respect to the quality of their mother. It is impossible for reasons of economy to keep the laying hens separate, a system of trap-nesting is resorted to, to make it possible to count the number of eggs laid by each hen. From the records of the Maine experiment-station it can be seen, that the number of eggs laid by an individual hen, is an unsatisfactory guide to her qualities as a breeder. It is obvious, that a system which would valuate the breeders according to the quality of their off-spring, would be vastly superior.

As all the pullets of a certain season have to be hatched within a short period, this means that the number of daughters one can hope to raise from one hen in a season is very limited, certainly below ten. A system of selecting poultry for egg-production which would be based on a comparison of the quality of groups of daughters of individual hens, would necessitate keeping the records of a very great number of small groups separate, and would be as impossible as the usual system of trap-nesting. On the other hand, a system based on a comparison of the groups of daughters from individual fathers is wholly practicable.

All that is necessary according to this system, is to hatch the eggs from each pen of one male and several females separately, and to house the pullets from these eggs in separate pens. This permits of counting the average number of eggs laid per hen in the progenies of individual fathers, and thus to select the group whose father had the most advantageous genotype. Sev-

13

eral generations of selection according to this system must lead to a rapid purification of the strain for the best genotype given in the potential variability of the original group.

This system of selection for egg-production in poultry is one of very few instances, where it is possible for us geneticians to point the way to the practical breeders. As we have pointed out several times, the practicians as a rule have worked out empirically a system of handling their material, which can hardly be improved upon by the geneticists.

———

SPECIES AND VARIETIES

ARE varieties incipient species, or is there any fundamental difference between species and varieties? A definition of the term species must cover what systematists havé been calling by that name and it is clearly inadmissible to use an old term for a new conception, especially if to do this we, have to limit the use of the term to a restricted group, a part of all. Our definition of species must cover such species as are known to be variable. Constancy as such, trueness to type, is clearly not essential. Lotsy has tried to give a definition of species by restricting this name for those groups of organisms, which are wholly pure for one genotype. We know that such species exist. Most of the populations of autogamous plants can be said to consist of a number of pure species, pure lines, and a few impure individuals. But to restrict the use of the term species for this special kind of species is as inadmissible as the restriction of the term dog to coach-dogs to admit of the simple statement that dogs are white, spotted all over with black dots. Such a description will never be true of that group of animals which are called dogs by everybody else, and Lotsy's definition does not fit the majority of groups called species by systematic zoölogists and botanists.

Nevertheless, species are strangely pure, and if a species does not necessarily consist of geno-typically identical individuals, the usual procedure, the description of a typical specimen as the specific type is assuredly founded on the observation,that an enormous majority of the plants or animals grouped under the name conform to the description. It is the current view among systematists that a species is stable, that the individuals belonging to it which are somewhat different from the type have no power to change this type, and that the existence of the aberration is merely a temporary one.

According to the opinion of the systematists, species are realities, real groups of organisms of which the majority are true to type, conforming to the description of the type-specimen. Species are thought to be unchanged, the species which existed as such long years ago are believed to be identical with their ancestors, and it is believed that if a number of species we now know, will not become extinct, they will be found unchanged after many years.

Species, according to the systematicians differ from each other in groups of characters, varieties on the other hand do not have the permanency of species. They differ from the species *to which they belong* in one striking point, and they are continually being produced by the species, they have no continuity in themselves.

A variety is a description of such individuals which are occasionally found within a species, which differ enough from the type to warrant a new and common name. Varieties do not commonly procreate themselves for any number of generations.

De Vries tried to find a genuine distinction between what constitutes species and what are varieties. According to him, species differ from each other in presence and absence of pangenes, whereas the difference between a species and its variety or between varieties would be due to different states of the same pangene, patency and latency. The result would be, that hybrids between varieties or between a species and a variety would "Mendelize," whereas species-hybrids would be stable, and would show no segregation of genes in following generations. We know now, that no such difference can be demonstrated. All the evidence points to it that all such differences as are inherited are always the result of differences in zenotype, of presence and absence of genes. And a biomechanical view of inheritance excludes states of patency or latency of genes. According to Darwin, species change under the influence of natural selection, and there is no fundamental difference between varieties and species. Varieties are incipinet species. He clearly demonstrated, that there is no fundamental difference between the points which distinguish species in nature, and those which differentiate breeds of domestic animals or strains of cultivated plants. Although we firmly agree with the last statement, we draw another conclusion, namely, that different breeds of domestic or cultivated plants are not varieties but species. An inkling of this specific difference between the main breeds of animals is shown by the breeders, who persistently look for many different wild species as the progenitors of these breeds, In another chapter we have tried to show, in what particulars the animals and plants under domestication are subjected to processes and circumstances different from those among which wild organisms live. The main circumstances which have differentiated domestic breeds are a heightening of the variability of the material by cross-breeding, by a taking up of gametes, genotypically different, an isolation, which is keeping the variability of the group from becoming equal to that of the original one, and reducing it, and often selection, which directs the process of automatic purification of the genotype. Varieties of plants and animals under domestication may quickly become species under the influence of strict isolation and selection, in nature varieties are continually being swallowed up by the species which produced them. In a species in nature, in which some individuals are impure, heterozygous, for a gene which has a visible influence on the development, occasional individuals are produced with a new recessive character. All those individuals together may be termed a variety, Var. niger, Var. alba. Such aberrant individuals do not constitute a new species. They differ from the parent-species in one or a few correlated characters. And again, the variability in a species may be sufficient to allow of the occasional production of an individual with a dominant character, developing from a zygote into which two genes got together, which in combination produce a marked influence and not alone. Without some isolation, such individuals do not found strains of their own type, they constitute a variety. Some groups of individuals are so constituted and situated that they constitute one "Paarungs-genossenschaft", one association within which all matings are possible. Such groups have a certain variability, a certain number of genes, for which not all the individuals are homozygous. We have called this variability the Total potential variability, and we have given the following definition of what constitutes a species.

A species is a *group of organisms* which is so situated and so constituted, that it tends automatically to reduce its total potential variability and which for this reason tends to become pure for one specific type.

We call a variety *those individuals together* which differ in some marked way from the common type, when there is nothing in these qualities or in the circumstances, which isolates these individuals from crossing freely with the typical ones.

It will be seen from these definitions, that apparent trivial differences in circumstance may decide whether one or two aberrant individuals will be specifically distinct or will be only a new variety.

In habitually self-fertilized organisms the course of evolution is fundamentally different from that in allogamous organisms. In autogamous plants there are no varieties.

In allogamous plants and animals the rule, that species differ in groups of characters and varieties in single ones is of common application. New species can only arise through some sort of isolation, of a group with a potential variability distinct from that of the type. In such a group the total potential variability gradually diminishes, and it becomes relatively pure for its own type, chance only deciding in how many points it will differ from the one or several species from which it originated. Varieties on the other hand are produced by chance combinations of gametes both lacking a certain gene, or supplementing each others genotype and so giving a new character.

Varieties can be given-off by a relatively pure species. In habitually self-fertilized plants every geno-typically different individual, is effectively isolated and protected from the swamping effect of random crossing. After every cross, such as occasionally even take place in barley or rice, a multitude of plants of divers genotype are produced, each one a potential species, each one rapidly becoming pure for its own genotype.

Natural selection must, if anywhere, affect the final qualities of a population of autogamous plants by a selection between the divers pure lines. It must be remembered, that a mixture of very many different equally well-fitted pure lines will gradually become poorer by the dropping out of those lines, which in a given generation do not happen to be included in the number of germinating seeds.

What should be the stand-point of the systematists? We have seen that they believe in the stability of species and that their whole procedure depends upon this relative stability. The alienation between systematists and geneticians has been mainly brought about by the acceptance by the latter of Darwin's hypothesis of the changes which species were thought to undergo by selection.

It is the very great merit of Wagner to have pointed out, that species are not changed, that they cannot be changed. I am firmly convinced that on this point Wagner has proved to be in the right and Darwin in the wrong. Darwin's observations on this point were made on cultivated plants and animals. These conclusions cannot be generalized, for in these organisms we find that one of the main factors in their change is an isolation, stricter than any found in nature. If species of cultivated plants and animals under domestication change by selection, as we have seen the Airedale terrier and the Collie changed, this change can easily be seen to be effected by a rigorous isolation, the breeders using only very few animals in every generation to continue the breed. It may be imagined that in exceptional circumstances an analogous process caused by an analogous chance combination of circumstances can occur in nature, but assuredly it is not the normal process. There is no reason for systematists to change their attitude in regard to the species-concept. If they have been estranged from Genetics by the fact, that they were in possession of evidence which showed species to be unchanged for long periods of time, it is to be hoped that their interest in Genetics can be revived and a better coöperation made possible. As to the stability of species, we have now to concede to the systematists the correctness of their view. Is the attitude of these observers toward species and varieties and their nature warranted by the facts, and is their method of applying names still adequate? We are convinced it is. Varieties cannot be truly said to be descended from species in the way in which species descend from each other, in so far as this implies that at a certain moment a variety begins life as such, by diverging from the main body. A variety has no necessary continuity through any number of generations and as the individuals which can be brought together under one varietal name, have their origin in chance combinations of gametes deficient in the same gene, or of gametes which bring into the zygote two genes, which only in combination have a definite action on the development, varieties usually differ from species to which they belong, in the effect of one gene less or one gene more. We must remember, that such variations between the members of a species as are due to presence and absence of genes having only a very slight influence on the development will commonly be classed as falling within the normal fluctuating variability of the species. A systematical description of a fauna or flora cannot be expected to be other than partially complete.

Only striking varieties, varieties in other words, which owe their difference from the type to a difference through presence or absence of a gene, which has a striking definite effect on the development, will receive a name. A difference in presence or presence of one such a gene will translate itself in a difference in one character mostly. Of course there are several instances in which a gene has a definite influence on more than one developmental process and therefore on several characters, such as the gene studied by Herbert Nilsson as influencing the colour of the veins in Oenothera. But here we are concerned with the average effect of the presence or absence of one gene. New species on the other hand, arise in a way, fundamentally different from varieties. A new species is a group of individuals originated from a limited number of plants or animals in some way isolated from the body of the species. Such a group must automatically become pure for its own type. The group can be said to constitute a new species, if the type is a new one, in other words if the total potential variability of the isolated group admitted of such a new set of genes. The formation of a new species out of one old one must be rare, the total potential variability of an old species not being sufficiently large. New species will mostly originate by isolation of a group of animals or plants belonging to a species, of which the total potential variability has been recently heightened by a cross (with some other sub-species). New species, will, with rare exceptions, differ from already existing ones in several characters, because the genotypic difference is one in several genes.

It follows from what is known about the action of genes, their distribution and stability, and about the causes for geno-variability and specific purity, that varieties differ from the species to which they belong and among each other in single characters, whereas species differ from each other in groups of characters.

Without the necessity of breeding-experiments or physiological tests being required to decide whether a few aberrant individuals constitute a variety or a species, we have a very simple morphological test, applicable to dried and flattend out herbarium-specimens and empty skins of animals. But far from being new, this criterion is a very old one. It is the common systematist's criterion of what constitutes a species and what a variety. To resume, species are realities, and they are stable, not changing. Further, we believe that those individuals which are seen to differ in one striking point only from the members of a species in the midst of which they live, constitute a variety, whereas individuals differing in a group of characters from hitherto described species — constitute a new species. In both cases systematists have for a long period had an opinion, differing radically from that of the geneticians, and the later genetic evidence all points to the fact that the systematicians have been right.

In this connection, the matter of denomination should receive some attention. It would seem, to a great many authors as if it were greatly a matter of personal taste whether a given group of specimens should be divided into two, or twenty, or two hundred species. What should be our standpoint? All the evidence goes to show, that species are realities, not only convenient groups made up at will. Therefore the number of species into which to divide a drawer of skins should be definite. The nature of the material should have a great influence. In autogamous plants excessive polymorphism is the rule. Here the very nature of the material makes species out of every type. Every plant is isolated from random crossing with others, and in a few generations its descendants will be all homozygous, will have a total potential variability — zero. Such material will therefore consist of a host of pure species. Here the "splitter of species" certainly is in his right absolutely. Nevertheless, the whole group, the combination of a whole group of species has certain qualities in common with one species in the allogamous organisms. If on a certain day there are fifteen hundred different species of oats, every one of these may have a total potential variability zero, but the whole of the fifteen-hundred combined have a very great Total potential variability, and this total potential variability has a tendency to reduce itself. In such a group of species new ones are constantly being produced as a result of occasional crossing, but on the other hand all the time some of the species are becoming extinct. To keep intact a collection of several hundred species of wheats, it is necessary, carefully to conserve a plant of each number for seed every year. This is not only necessary to keep the species under name and record, but it is essential for keeping the full assortment intact. If seeds of all the different species were mixed and sown and harvested promiscuously, not every species would happen to be represented in the quantity of seed saved from the harvest, which would be enough to grow an equally big field next year.

The point we want to emphasize is, that, whereas in wheat there are an enormous number of real, concrete species, there exists an abstraction, the combination of such species, wheat, Triticum vulgare, which has its own total potential variability and tends to reduce this automatically.

In the matter of nomenclature, if we want to use the name Triticum vulgare for wheat, we cannot use such names as Triticum miracle or Triticum red fife for the species, they are not of the same class, and certainly not of the same order as Triticum repens. The logical terminology here is a trinominal system. Triticum vulgare can be used as the name of the whole group and the component species can be properly called Triticum vulgare miracle and Triticum vulgare squarehead. If we only remember that such things as Triticum vulgare are not species but combinations of species, no harm is done. The different wheats are certainly not varieties of Triticum vulgare.

A trinominal system may look cumbersome at first sight, but it makes it easier to denominate very many closely related real species. To conclude, we would once more state it to be our opinion that there is a fundamental difference between varieties and species, and that only under rare, peculiar circumstances, but which are often realized under cultivation, can varieties become species.

THE LAW OF JOHANNSEN.

THE way in which an organism develops, determines its qualities, and in the end the behaviour of an organism under the opportunities given by its environment depends upon the reaction of its cells upon their immediate environment. This reaction of cells, in so far that it is different from that of cells of other organisms, is dependent upon their constitution, chemical and physical. And ultimately, we think of this constitution of the cells, as given in the set of genes inherited in the original cell or cell-complex from which the organism grew up.

It is evident, that the difference in reaction, which we observe between groups of cells in one and the same individual, cannot be expressed in terms of presence and absence of genes, for no such fundamental difference between the cells of one organism is observable in those cases in which we can make a somatic cell reproduce a whole individual.

Nevertheless the difference in reaction upon the immediate environment, which we observe between cells and cell-complexes within one organism must be due to a difference in the constitution of these cells.

The constitution of a cell cannot be determined by the mere presence of a definite set of genes, heritable substances, for if it were, the difference in quality between cells of one individual, or of one pure clone of uni-cellular creatures would be accompanied either by a change in the set of genes, or by a change in the quality of the genes themselves.

The conception of genes as direct determinants for qualities necessitates the assumption of a qualitative instability of the genes themselves. And the fact that we have been trying to account for the facts of variability without assuming a qualitative instability of the genes, must be explained by the wish to try out to the full a biomechanic theory of inheritance and life rather than to fall back upon a vitalistic one, which must be as sterile as on the face of it may seems satisfactory.

The hypothesis, that the genes are relatively simple substances with autokatalytic properties, has the advantage over a vitalistic conception of genes, that it is a hypothesis which can be worked with, even though it may not be found to work in all cases.

The question as to the fundamental qualitative stability of the genes themselves is to my mind of the very first importance, and all our work of the last years has been planned so as to shed light on this point. It is indeed remarkable to observe the casual way in which Geneticists who themselves perform experiments, the results of which have a very direct relation to the subject, refer to the problem. Vitalism has a very strong grip upon the minds of most of the Geneticians.

If we observe great differences between cells and cell-complexes in their reaction upon the direct environment, and we have to admit that nevertheless the personnel of the genes is identical in these cells, must we conclude that such differences are due to the inclusion in some substances or physical relationships as foreign to the protoplasm of the other cells as a vital dye, or as the effects of a pressure between two plates of glass?

If we use the hypothesis, that the genes are not in themselves living, but chemical things, we can conceive of protoplasm as a combination of substances which each have the property of being a katalyzer for its own synthesis, and which can therefore reproduce themselves quantitatively without changing qualitatively. And we can imagine how this combination, protoplasm, derives its physical properties from inter-relations between several of its constituents. In other words, it is wholly unnecessary to conceive of protoplasm as of something distinct from, and including genes, but it is both simpler and better in accord with all the facts to conceive of protoplasm as of the sum total of the genes.

It is entirely feasible to imagine combinations of autokatalytical substances which have no structure, and no "body",
but which nevertheless, when occurring together manifest
their presence in a way which is different from the action of
each alone. (Filterable virus). Causes which destroy the physical relationship between the constituents of protoplasm may
cause death, that is, the relationship as such may be self-con-
tinuing, and destruction may be irreversible. On the other
hand physico-chemical structural relationships are not necessarily dependent upon combinations of several constituents.
The formation of a crystal in a super-saturated solution is an
autokatalytical process in which the structure can be irrepar-
ably destroyed.

The conception of living matter as a combination of autokatalytical constituents makes the essential thing in living
independent of its more usual manifestations. A small, dry
crystal in a closed vial in a cold environment, is as full of latent
life in the narrowest sense as a dry seed under the same circumstances.

A combination of even two things can have a set of qualities
of its own, quite different from those of its constituents, but it
is evident that the continued reproduction of both constituents is essential for the continued existence of the combination.

Variations in the proportion in which the algae and the fung-
us mycelium occur in the combination, which we call a lichen
will produce variations in the properties of the lichen.

A fragment of a lichen may for a time reproduce its qualities, in as far as they are dependent upon a certain proportion
of the constituents. But it is clear that a small mass of the
lichen, will have the same potential possibilities of differentiation, only so long as it includes both constituents.

In the same way, the possibilities of a cell, or of a series of
cells, are plainly given in the different relative quantitative
proportions in which its constituents can occur.

Within the limits of a continuation of life, one or several
cell-constituents can become predominant in the combination,

if the circumstances favour such a predominance. Such quantitative changes in the proportion of the substances which make up protoplasm, are transmitted from cell to cell. And it becomes evident that, in those cases where the whole organism is but one cell, this transmittance of definite proportions between the substances which make up protoplasm is in a great measure "inheritance".

The proportion in which the constituents occur, is greatly a matter of circumstances, and changes in this proportion are mostly reversible, though not necessarily so.

This potential reveisibility of changes of proportion between cell-constituents is in a very sharp contrast to the irreversability of another process, namely the loss and the (at least hypothetically thinkable) acquisition of a gene.

It appears at first sight as if the processes, which change the equilibrium in the proportion between the autokatalytical constituents of protoplasm, could easily bring about the complete exclusion of one or several of these constituents, genes.

Why is loss-mutation such an exceedingly rare phenomenon? Thousands of genes must be present in the higher organisms, to judge from the number which we can study in some, whereas we know that such a study becomes possible only through comparison of individuals with and without, and is therefore limited to a great extent to relatively vitally unimportant ones. But loss of a gene as distinguished from absence from a new combination, caused by redistribution, which is commonly confused with loss-mutation, is a phenomenon so rare as to be certainly negligable as a factor in evolution.

As we will see later, the facts point to it that within the nucleus the original set of genes, such as it is inherited by the original zygote, is conserved intact. The great diversity of function and of gross chemical constitution of the cell-complexes within one organism, the facts of functional adaptation admit of a simple explanation on the hypothesis that the genes have autokatalytical properties. It is evident however, that the relative quantitative preponderance of genes which we

look upon as causing this diversity, is limited to the protoplasm outside the nucleus, to the cytoplasm exclusively.

All the recent observations on the effect of selection in
clones of uni-cellulars point to this same conclusion, namely,
that there is a very great difference between an inheritance, a
transmission of substances, genes, through the cytoplasm on
one side, and through the nucleus on the other. We know that
there is a list of cases of more or less pronounced maternal
inheritance. In some of these cases the mechanism is clear,
such as in the cases of infectious spotting in plants, and even
in the cases of the transmission of self-propagated corpuscles
outside the nucleus. In other cases the preponderance in bulk
of the cytoplasm in the gamete derived from the mother,
makes the explanation simple.

It is certainly significant, that we are gradually eliminating
the alleged instances of non-Mendelian inheritance, and that
these instances are rapidly being restricted to cases of a continuation of cytoplasmic conditions through germ-cells, or
through whole embryos.

A new distribution of the genes over the nuclei, occurs almost exclusively, at cell-divisions preparing for sexual reproduction. Exceptions are found, such as instances of somatic
segregation in pure clones (Salaman), but they can generally
be recognized as such.

So long as there is no such process in a given material,
we can say that there is no genetic variability in it. Ordinarily
there is no genetic variability in pure clones, in groups which
have been derived by exclusive vegetative reproduction from
one zygote. And there is no genetic variability in what Johannsen has called a pure line, in material which is derived by
exclusive self-fertilization from one homozygous plant. In such
material by definition all gametes produced are identical, and
all zygotes must be, so that (with the exception of the case of
a spontaneous loss) all the nuclei in such a material contain
the same set of genes.

Johannsen observed, that selection in material which was

pure in respect to its set of genes, did not produce any heritable change in this material.

From these and later observations, we formulated what we called the law of Johannsen: "The nature of the genes does not admit of qualitative variation" (1915).

We will have to examine this law in the light of some recent papers which show the effect of selection in pure clones of unicellular organisms.

Johannsen's law of the qualitative stability of the genes is derived from repeated observations as to the purity and stability of all those groups of organisms, which we have good reason to believe to be identical in respect to the set of genes carried, a purity and a stability which is not permanently affected by a change in the environment or by selection.

Whenever we want to discuss this law and its universal validity, we must remember that in its simplest form, as we have formulated it, it only states that the genes themselves are qualitatively stable. The usual ineffectiveness of selection in pure lines and pure clones for instance, we explain by assuming that in such material the individuals really have all the same set of genes. But at the same time we know, that the ineffectiveness of selection and of change of environment to change such material, is only the result of this nature of the genes, and will hold good only so long as within this material there do not occur processes, which change the set of genes carried. And we must not confound a temporary change of the quantitative increase or decrease of genes in the cytoplasing a number of cell-generations with the qualitative change in the genes themselves.

We may certainly not turn round the statement that genes are qualitatively stable, and that therefore, the germ-plasm of material pure for its set of genes is not amenable to change by selection, and make it read so, that the test of the effectiveness of selection in a pure line or a pure clone is a test of the validity or otherwise, of Johannsen's law.

A pure line is a group of individuals which have arisen by

self-fertilization out of one homozygous individual. A pure clone is a group of individuals which originates by an asexual process from one individual. Now, ordinarily, changes in the genotype of individuals of a pure line are rare indeed, and so are corresponding changes in the set of genes carried by the members of a pure clone. And therefore, ordinarily selection within a pure line or a pure clone, must be ineffective. But there are certainly exceptions to the rule, that selection in pure lines and in pure clones is ineffective, simply because there are instances in which within pure lines quite apart from any selection, quasi-spontaneous changes in the set of genes transmitted have been noticed. And a little more frequently, analogous changes have been noted in pure clones.

Johannsen has observed such a mutation, the spontaneous loss of a gene from a bud in beans. Nilsson Ehle has observed a few instances of real loss-mutation in oats and wheat. In clones, we occasionally meet instances of a kind of vegetative segregation. We find, that an individual heterozygous for a certain gene may produce buds in which this gene is lacking. Very many horticultural sorts of Dahlia, Chrysanthemum and Azalea have originated in this way. That we are here concerned with a vegetative segregation and not ordinarily with loss-mutation, comes to light in those instances in which a plant which in an asexual way produced individuals unlike itself, is tested by self-fertilization. The results of Salaman with potatoes are very striking. Some plants with elongated tubers produced occasional round tubers, which in their turn would produce only round ones. Self-fertilized, the plants with the elongated tubers, gave a minority of off-spring with round tubers, thus showing their heterozygous nature.

A good way to bring out the relation between the law of Johannsen and the effect of selection, is to compare these effects of selection, to the behaviour of falling bodies, and the relation between this behaviour and Newton's law of gravity. The attraction between bodies is in direct proportion to their mass and in inverse proportion to the square of the distance between them. This is the law, comparable to our law of Johannsen.

One of the results is, that in the absence of the resistance of air, in the ideal conditions of a vacuum, bodies fall with equal velocity. We may compare this result under ideal conditions, which illustrates the law, to the fact, that selection is ineffective where the selected material remains homogeneous in respect to the genes present for the duration of the selection-experiment. In everyday life we observe, that the velocity with which bodies fall, is influenced by their specific gravity, and we observe bodies which do not fall at all, or fall away from the earth. And these observations, which at first sight look incompatible with the experiments under ideal conditions,' may be compared to the observation which we can make every day, that selection changes the mode of nearly every character selected.

In the field of Physics, this difference between Newton's law and the experiences of everyday life is certainly felt as uncomfortable by beginners. But the fashion just now is all against little experiments showing, that in a more or less complete vacuum it is possible by accurate measurements to see certain objects always arrive at the bottom a little ahead of certain others. Not so in Genetics. We have to proceed with the utmost caution. If we say, that the non-effectiveness of selection in pure clones proves the qualitative stability of genes, the result of selection-experiments within pure clones are shown us, in which indubitable changes have been brought about.

We think there is a tendency on the part of many Geneticians to see the importance of the work of Johannsen and others with pure lines, in the direct application to our conception of the nature of species, rather than in the light it sheds on the genes. Johannsen's conception of the existence of pure lines is frequently by his followers, but more often by his critics, exaggerated into a pure-line theory of evolution. We think, that his work has a very important, but decidedly indirect

bearing upon the evolution question. Through selection-experiments with pure lines there has been discovered what we
have called "Johannsen's law," which states that the genes
themselves are qualitatively stable. The validity or otherwise
of this law, is certainly of the utmost importance for a solution
of the evolution-problem. For, if we need not take into account
the change of the genes themselves for evolutionary processes,
we can limit our research to those processes which are compatible with this qualitative stability. For one thing, this does
not exclude selection. We know, that ordinarily selection does
not affect a pure line, but this does not mean that it may not
affect other groupings of individuals. The work with pure lines
has taught us valuable things about the nature of genes, and
incidentally about pure lines.

Johannsen showed, that many at first-sight pure species, in
reality consisted of mixtures of pure lines. But it does not
seem logical, that those authors who have accepted as proved
the law of the stability of genes, as demonstrated by the ineffectiveness of selection in material which is geno-typically pure,
and so long as it is pure, should continually be exposed to crit-
icisms which are directed against their supposed view, that
all species are composed of stable, pure lines. So far as we know,
the only author who has proposed a theory of evolution based
upon this supposition, is Lotsy.

The existence or non-existence in nature of "pure lines" is
immaterial to the law of Johannsen. We happen to know, that
in certain organisms such pure lines do occur in nature, but
even if they would not have eixsted in nature, it would have
been possible to discover and formulate the law. Even in
those plants where no pure lines exist, selection-experiments
in artificially self-fertilized series would show a diminishing
effectiveness of selection, going parallel with homozygosis,
and it would have been possible to deduce Johannsen's law
from such experiments.

The best demonstration of Newton's law is the experiment
in which diverse bodies are made to fall in a vacuum. The fact

that arround us a complete vacuum is not met with, does not invalidate Newton's law. And the view that we, who accept the qualitative stability of genes must therefore necessarily hold the fallacious theory that species in nature are pure lines, or are built up of them, is as absurd as the view that a physicist who accepts Newton's law must necessarily believe in a universal vacuum on earth.

Selection in a pure line could theoretically result in a change. This does not mean that selection can change a gene, but that within a pure line genotypic changes are possible, which might furnish the material for such an effect of selection.

Selection in a pure clone can also result in a change. The fact that so far no effect of selection within pure lines has been noted, whereas it has been observed in clones, is only natural if we remember that mutation in homozygous material means loss of a gene, wheras spontaneous geno-variation in a pure clone can be vegetative segregation, a production of cells lacking in a gene, by cells heterozygous for it.

It is conceivable that pure clones may at the same time be pure lines, that is, clones may be started from one homozygous individual, just as pure lines. Such clones would presumably be as proof against change by selection as pure lines. But the very fact that vegetative segregation is not comparable with loss-mutation, must warn us against getting mixed up in our terminology. We certainly may not use the term pure line indiscriminately for pure clones. The fact, that we can see vegetative segregation in a group of plants cut from a single heterozygote as well as in the original heterozygote disturb the more usual uniformity, gives us sufficient reason to look for this process as an additional source of hereditable variation, added to mutation.

It becomes more and more clear, that the ordinary heritable differences between organisms are caused by underlying differences in the set of genes present in their cells, differences to be expressed in presence and absence of individual genes. The permanence of the differences which we observe between ge-

netically pure groups, we think of as caused by the fact that
absent genes are not spontaneously acquired, and that reversely,
the loss of a gene, for which an organism is pure, is a
rather uncommon process. But apart from such permanent
differences as due to presence and absence of genes, can we not
also conceive of more transient differences due to quantitative
inequalities between the amount of a gene present in two organisms which both have? At the outset, we must try to make
it understood, that quantitative differences in the amount of a
gene coöperating to the development are fundamentally distinct from qualitative changes in the genes themselves.

Two bronzes, both composed of copper, zinc and nickel may
be very different in very many qualities according to the relative quantity of each of the composing metals, but there is a
fundamental difference between this relative proportion as a
cause for the diverse quality of the bronzes and a hypothetical
change in the nature of the copper or of the zinc. In fact, such
a qualitative change of the copper is not warranted by the
facts.

The view, that there is a possibility of differences between
cells caused by a more or less of certain genes, is bound up with
the conception of genes as essentially chemical rather than
morphological. There is no other logical way to reconcile the
patent difference between cells of one individual with the genotypic sameness inferred from diverse observations. The only
alternative is Weismann's idea of a gradual simplification of
the genotype in generations of cells throughout development.
The facts of regeneration, of vegetative reproduction from one
epidermis-cell of certain plants, facts which have come to light
in a study of graft-hybrids, all contradict such a gradual dropping out of "determinants".

If we conceive of protoplasm as of an emulsion in which
genes play an integral part, it becomes clear that, without a
change in the personnel of genes present, great differences between cells and tissues made up of cells can be due to relative
proponderance of one or several genes, just as great differences

between alloys can be due to relative preponderance of one or several constituent metals. And therefore, such differences between cells can be thought of as ultimately due in a great measure to the "environment" of the cells in as far as it furnishes the "ingredients" for some genes in excess. If we conceive of the genes as of chemical substances with autokatalytic properties, rather than as of "vital" protoplasmic "determinants," we might take as a simple illustration any relatively little complex autokatalytic substance, such as oxide of iron, and imagine it for a moment to be a gene and a factor in the development of some organism. We will see how a few generations of cells in an evironment with plenty of iron and oxygen, have come to contain a great quantity of this gene, FeO_2. The presence of much of this gene may well become appreciable as a colour, let us say red. Now, even if the cause of this redness changes, if the circumstances which brought the cells into contact with iron and oxygen, alter the daughter-cells of a red mother-cell will be redder than the average. No matter how abrupt this change will be, the change in characters, colour through cell-generations will be comparatively gradual.

If, by some uncommon constitution of the environment, a sea-urchin egg shows very dark colour, its two, its four blastomeres, will yet be redder, than the common run of cells in twocell and four-cell stages. We might measure the coëfficient of correlation and find it very high between a mother-cell and its daughter-cells.

Now, if we remember that those daughter-cells a moment ago were the mother-cell, we begin to ask ourselves the question: How does this resemblance between mother-cell and daughter-cell, between egg and blastomeres in a sea-urchin compare to the resemblance between a mother-sea-urchin and a daughter-sea-urchin?

The process of transition between a mother-cell and its two daughter-cells is so direct, that all the factors in the development of the mother-cell, genes as well as non-genetic factors, are also factors in that of the daughter-cells. The quality of a cell is the result of its genotype, and of the way in which it has

reacted upon its environment, and as daughter-cells are but
halves of it, their characters are those of the mother-cell. Not
entirely, because the influences to which daughter-cells are submitted from the moment of division up to the moment at
which they are themselves ready for division, may be different
from those under which the mother-cell completed growth.

In the development of metazon, each individual has not
been one half of its parent up to the moment at which it had
half its bulk. In the ressemblance between parent and off-spring
we have here to reckon far more with the genotype, with the
set of genes present.

In so far as the set of genes of the daughter and that of her
mother are alike, the reaction upon the environment will be
alike, and in so far will they in the long run show likeness: in so
far as their genotype is unlike, they will in the long run show
difference, provided the difference is one, which affects a gene
which has in this type an effect upon development.

Peculiar conditions, which pertain for a period of a few cell-
generations have no appreciable influence upon the final qualities. The final qualities of the mother are the result of her development from one cell to a complex organism under the influence of her genotype which remains the same, and of her environment which may fluctuate. The final characters of the daughter are also caused by her genotype which remains the same
and the environment which may fluctuate. The longer the
period of development relatively, the more these transient
fluctuations in conditions are equalized in the long run, the
more therefore likeness becomes a true test for sameness
of genotype.

We may compare the qualities of an adult brown bear,
which has grown up through several dry and wet seasons,
through fat and lean days, passed through several hibernations
to the qualities of an adult dauhgter of her age and experience,
and observe a certain likeness as compared to a certain unlikeness between these two on one hand and a pair of Malayan
bears on the other hand.

This likeness we consider a better evidence of sameness of genotype, than the likeness between two summer-generations of a certain Daphnia as compared to the unlikeness between these two on one hand and winter-generations of a Daphnia on the other. And in my opinion, the comparative likeness between a mother-cell and a daughter-cell is certainly not of the same order as that between a mother bear and her daughter.

If we simply speak of inheritance of characters, we mean quite a different thing according to the type of organism under discussion, and we must not forget this, or confusion will be the result. One man may happen to work with wheats and "inheritance of characters" may mean something definite to him. It is possible to show him a definition, which he will think acceptable, and then show him an instance, which complies with the definition but not with his conception of "inheritance of characters". Simply because, in accepting the definition, he has his own material in mind, and he has not quite thought out what "inheritance of characters" might mean in other material.

The "characters" of any individual are its qualities, nothing more nor less, and those qualities are simply the result of the development, the way in which the individual grew to be from what it was before. Now in the bigger, longer-lived, multi-cellular organisms, the "environment" is relatively so constant, and variations in it so tend to counteract each others effect, that in the long-run environment will be on the average about the same for every individual. And in the measure in which this is the case, environmental factors in the development become more and more modification-factors, and less and less determination-factors, to use the terminology of Roux. Now, the smaller the organism, and the smaller the number of cells composing it, the more will environmental factors in the development become determination-factors.

If we are used to work with wheats, where we compare a mother-plant which has grown from September till July with a daughter-plant that has grown from September to July, the effect of a difference in the environmental factors will appear altogether negligible to us, and if we study the inheritance of characters, we study genotype differences, and we are more or less liable to call these differences, "unit-characters". In this material, recurrent, characteristic differences, unit-characters, are the obvious and direct result of presence or absence of genes. And if we dislike this term "unitcharacter" with its suggestion of Weismann's "determinanats," the dislike is born of our biomechanic conception of characters, and not of any unfitness of the term "unit-characters" for describing these differences.

If we say "heredity of characters," we think of genotypic likenesses between the individuals having the same heredity, although the term implies nothing of the sort. The red colour of Sudan III, transmitted to eggs and larvae from moths having eaten it, in Gage's experiments, is just as much an inheritance of a character as the inheritance of red colour in wheat, where we know one of the genes, is an inheritance of a character.

Or is it not? And it we shrink from calling both processes by the term inheritance of characters, which is a matter of personal taste, should we call the likeness between a mother and her daughters in uni-cellular organisms inheritance of characters? To my way of thinking, most of the confusion, most of the startlingness of certain selection-experiments, most of the heated controversy round about Johannsen's work, is caused by loose thinking and unwarrantable generalisation of facts found in wheat to yeast, and of conditions found in bacteria to what they are, or are not, in peas.

We must remember, that the likeness between two cells, of which one is half of the other, or has been, only half an hour ago, as compared to the unlikeness between these cells and a descendant cell fifty generations hence, is much more influenced by a greater likeness or unlikeness of the environment, than the likeness between two adult hens, mother and daughter.

No one, we should think, likes to call "heredity" of characters" the likeness of a tissue-cell and its two halves, daughters,
and compares this likeness to the likeness between two hens.
But when the separate cells are free-living individuals, they
become in our opinion so much more like hens, that we are apt
to overlook, that here the likeness between mother and daughter is not of the same order.

However, we meet in higher animals and plants another
sort of likeness, which is directly comparable to the likeness
which exists between mother-cell and daughter-cell, rather than
between cells which have only the same genotype.

The experiments of the Viennesse school of Lamarckians
have abundantly shown, that the likeness between parent and
off-spring, who both grew up changed by the same peculiar
influence, can be interpreted as inheritance. We hear that mice,
grown up in high temperature have long tails, and that their
children, if they grow up in the same environment, also have
long tails. We are told, that a certain viviparous lizard changes
its belly-colour from white to red in high temperature, and that
the young are born with a red belly, even if the mother, just
before giving birth to them, is put in a moderately warm place.
And we hear such instances, cited as cases of inheritance of
acquired characters. We are apt to make light of such an
interpretation and to say that it is no more remarkable, that
the young are born red in a hot environment, than that they
are born hot. But why should we object so much to seeing such
instances brought forth as cases of the transmittance of acquired characters)? Simply because we suspect these Lamarckians
more or less injustly, of meaning characters in the sense of the
"unit-characters" of the Weismannian Mendelians, and of thinking, that they have proved an inheritable genotypic difference, induced by the high temperature. Our experience with
Lamarckians is, that they as a rule do not want to attribute
such a meaning to their experiments at all, even though they
are not adverse to seeing other naturalists rashly jump to such
a conclusion.

After all, characters in Metazoa are never transmitted, only genes and combinations of genes are, and it is not more, nor less wrong, to speak of the inheritance of acquired characters, than it is to speak of the inheritance of innate characters.

Now, whereas there is no direct transmission of characters in more complex metazoa, there certainly is in all uni-cellular organisms.

If we notice a peculiar spine-like excrescence in Paramoecium such as observed by Jennings, we may note how it gradually diminishes in size and eventually disappears, and think nothing further of it. And even if we see, that during this process of disappearence of the spine at one end of the organism, the other end of the animal is four times lost and regenerated, we may not see any important connection between these two sets of facts. It is more or less a matter of temperament. We may now look upon this end which carries the excrescence as upon its own daughter, for the whole animal divided into two and we call the halves daughters of the individual, which we had in our hanging drop before. And therefore, the end with the spine is not only the same individual from the time at which we began to observe it, until the moment at which the spine was lost, but also the great-great-grand-daughter of the original animal, and therefore of itself, and if we so please, we certainly can say that this character, the spine, is inherited through four generations.

Only, we must be very clear, that this inheritance of this spine is a different thing from the inheritance of a tail through four generations of chickens. For the egg has no tail, and this organ is made anew by the young individual at the stage of development at which its mother grew a tail. The tail did not merely persist like the excrescence on the Paramoecium, or the virulence of the bacteria, or the Sudan III in the moths. Its development is a function of the constitution of the generations of cells which link the daughter to the mother, in the sense that the presence of a tail in both, as compared to the possible absence in other chickens, is more certainly caused by a like difference of their genotype from the set of genes of the tail-less individual.

We must see, that the similarity in size between a cell and its daughter-cell, and the similarity in size and spininess of the shell around those cells in uni-cellular organisms, and all sameness of this nature is merely a sameness of the characters which are for both cells the result of a coöperation of the same genes, and the same non-genetic developmental characters. There is no room for any great unlikeness through change in the non-inherited developmental factors, through change in anything but the genes. The similarity between a shell of Difflugia and that of its "daughter" is not of the same order as the similarity of two pea-plants of the same pure line, not even of the same order as the similarity between two trees budded from the same "mother-tree." And for the same reasons the distinction between a clone of Difflugia with large shells and one with small shells, is not of the same order, as the difference between a strain of mice with long tails and one with short tails, and not even of the same order as the difference between a clone of potatoes with large leaves and one with small leaves. The difference between the clones of Difflugia need not be due to genotypic difference at all, whereas the dissimilarity between two clones of potatoes is almost certainly due to a genotypic difference.

If we take a dozen buds from one tree and graft them on a dozen seedlings, and we observe that the twelve trees grow to the same height, this similarity in height shows similarity of genotype. If we split a certain tree in halves length-wise and succeed in making both halves live, the similarity in height is not of the same order as that, which we observed in our clone of twelve budded trees. If we carefully split a big tree in winter and count the number of leaves which each half produces in spring, we can calculate the coëfficient of correlation between these numbers as compared to the average number of leaves on this sort of tree. This would be a piece of work directly comparable with the studies on Difflugia of Jennings (and similar work on uni-cellulars). What relation do these experiments have to Johannsen's work with pure lines of beans and peas and so, and especially, are the results compatible with what we have called Johannsen's law, with the qualitative stability of genes? At first sight there seems to be no relation, but evidently some of the authors think their results to have a direct bearing upon the problems touched upon by Johannsen. F. M. Root says in his paper: The main problem, undoubtedly, is the one which has already been discussed. Are "pure lines" really pure? Do heritable variations occur within the clone? It is this problem which is attacked in the present paper." And in his conclusions: "The further idea, that within one of these "pure lines' no variation in genetic constitution is possible, except by a sudden mutation, large or small, is not direct observation but hypothesis". "But of late the tide seems to be turning somewhat".

We certainly admit that the idea, that no variation in genetic constitution is possible within pure lines, is a hypothesis. That is to say, the hypothetical part of the statement is the assumption that this absence of variation, and ineffectiveness of selection within such material is due to the fact, that every individual is pure for its genotype, which is the same as that of all the others. The absence of variation and ineffectiveness of selection of real pure lines is fact, observation, and not hypothesis. Before we discuss again the effectiveness of selection in clones and pure lines it is well to repeat, that there is only one instance recorded in literature, and this is the instance of Castle's interpretation of Hoshino's work on the flowering-time of the pea. We say Castle's interpretation, because Hoshino does not himself, find evidences for the effectiveness of selection in his results with pure lines.

Now the only point on which we beg to differ with Root and Jennings is, that we do not see that "the problem attacked in the paper" is really that which Root states in the words: "Are pure lines really pure" Some authors mix up the term "clone" and "pure line." Root certainly does not, but the fact that he thinks that selection-experiments with clones of uni-cellular creatures bear upon the problem of the purity of pure lines, shows, that he sees the point of the discussion in something different from what we consider essential. To our mind, the purity or otherwise, of clones and pure lines alike, is non-essential as compared to the qualitative purity or otherwise of the genes.

This is the real problem with which we are concerned. Are the genes really qualitatively stable? Have we to explain this purity of groups of individuals which originate out of repeated self-fertilizations, under all sorts of conditions, even selection, as due to a qualitative stability of the genes for which they are homozygous?

And if we state the problem in this way, some facts become very significant. In the first place we saw that no instances have been adduced, which show the effect of selection in mul-ti-cellular organisms in material which we have good reason to believe to be pure and homozygous. We are now leaving on one side Castle's example, as Castle has recently stated his con-conversion to a belief in the stability of the genes.

In the second place, we know that selection within clones of higher plants like sugar-cane and dandelion and beets and potatoes is ineffective with some significant exceptions, namely with the exception of the occasional quasi-spontaneous production of novelties. We know, that these novelties arise in some clones independently of selection, and if they occur during a selection, are more often in quite irrelevant direct-ions than not.

We further know from Salaman's work that such novelties produced by clones, are the same novelties the plants will produce by self-fertilization. In other words, these cases can be explained by somatic segregation, and they have nothing to do with qualitative changes in genes. We further know how in graft-hybrids disturbances in the arrangement of the genetically different cell-layers are often noticed, within the individual as well as during asexual multiplication. The work of Stout with Coleus rurnishes evident examples of both processes.

And next come all the facts, which show changes in the characters of series of organisms, which are obviously not dependent upon rearrangement of the personnel of the genes, but appear to take place in geno-typically pure material as well as in impure. Do they tend to make us believe in a qualitative change in the genes themselves?

The facts in this group are a rather miscellaneous collection, bien étonnés de se trouver ensemble. On one end come the experiments with the inheritance of red colour produced by ingestion of a red dye. We can see how this dye is handed on in pure material as well as in hybrids. But would anybody suppose such a transmission of a character to be due to qualitative changes in a gene? Next come the experiments performed at the Viennese Vivarium, by Przibram and Kammerer, in which mice with long tails as the result of life in a hot environment have long-tailed off-spring in the same hot room, and the transmission of a changed colour in lizards already discussed. And we think, that the experiments of Root and Jennings and Middleton belong to this class, because the material is of such a nature as to make probable the direct transmission of properties of cells, both properties different from the mode through action of the environment, and properties induced by quantitative fluctuations in the amount of individual genes present.

As we said above, there is no "pure line theory." And the pure line conception has no connection with evolution, at least no direct one. Root in his paper on Contropyxis sees two alternatives to bring his experiments into line with the "pure line" hypothesis. In the first place he thinks, it is possible to explain every instance of the effect of selection within pure clones of uni-cellulars by an individual explanation, like Morgan's attempt to explain inheritance of variations by assuming somatic segregations at the cell-divisions. In the second place Root thinks, we might assert, that the pure line concept is correct only on the average, and not holding for individual cases.

We have tried to show, that there is a third course, that quite apart from any existence of pure lines in one species or the other, the work with pure lines made us discover Johannsen in law, which has as little to do with pure lines as Newton's law with empty spaces, and which simply states that the genes are qualitatively stable. And we saw, that until now no facts have been brought to light, which are not in accordance with this law. We have seen how and when selection may even become effective in pure lines, but why such an occurrence has not yet been noted, and what is the fundamental difference between pure lines and clones, and on the other hand between metazoa and protozoa. We certainly cannot compare results in pure lines of metazoa with results in clones of protozoa, or state that the results obtained in one group cannot be valid, because they are not supported by results obtained in the other group.

We may not compare the inheritance of characters in higher plants and animals, where such "inherited" characters are produced anew by the children, to inheritance of characters like the red colour of a dye or any character of uni-cellular organisms, where the children have the character of the parent because a moment before we observed them as separate individuals, they constituted this parent.

Selection-experiments with pure clones would be very interesting and we would suggest to anyone, who entertains a doubt about the law of Johannsen, to start series of selection-experiments with pure lines and pure clones. We would think that only the very best material would be good enough for such experiments. The pureness of "pure lines" is of double origin, made up of homozygosis for the genes present, and of qualitative purity of the genes. Therefore, absolute proof of homozygosis of the material is necessary, to decide for or against the law of Johannsen. Loss-mutations should be watched for, but homozygosis is more important than anything else. We have seen that it is possible to show, that the viable seeds produced from unfertilized female flowers of squashes, are grown out of real fertilizable germ-cells. Hoterozygous plants produce unfertilized seeds, which by segregation show to develop out of real gametes. Such seeds must necessarily produce homozygous individuals, and the very pure lot of plants produced from one such an individual is theoretically the only really dependable "pure line" according to Johannsen's terminology.

On the other hand, experiments with clones should be undertaken with an understanding of what somatic segregation may do to the material, if the members of a clone are heterozygous. There are many plants, that could be chosen as material. If one starts selection-experiments in a clone, which starts from a homozygous plant, he may be sure, that he has now only loss-mutation to watch out for in an interpretation of his results. Wheat may be propagated asexually, and so may barley, and beans can be propagated by cuttings. It should not be difficult to devise a means of propagating such plants as the Cupid sweet-pea vegetatively. The results would be more free from criticism than those of any selection experiments in pure lines or in clones, that we have knowledge of.

EVOLUTION IN NATURE AND UNDER DOMESTICATION.

THE facts observed in animals and plants under domestic ation are of great interest for a study of the causes of variation, and of the effect of selection on species. But, before we can draw any far-reaching conclusions from these facts, we must examine the difference between what happens in nature and under cultivation. For, only if we know in how far conditions are similar, and in what particulars they are dissimilar, can we begin to generalize from the facts observed.

In animals and allogamous plants, a species consists of a multitude of more or less similar individuals, a "Paarungsge-nossenschaft," within which matings are free and inter-crossing is the rule. If nearly all the individuals have a given genotype, and therefore a certain set of characters, the very few aberrant individuals have no chance of propagating their type. Every aberrant individual mates with a normal, and those of its off-spring which are not pure for the common genotype again mate with normal individuals. The very existence of a multitude of genotypically identical individuals conserves the type of the species. Selection within a species, natural selection, has been exercising whatever influence it has for as long as the species has been living in the circumstances in which we find it. A group of animals or plants, impure for genes having a marked effect on the success in life of the individuals, will probably become pure for a certain genotype quicker than when no selection discriminated between individuals. If we observe variation within a wild species, the variability is most marked in non-essential characters.

Cross-breeding in nature, crossing between members of differ ent species, heightens the variability of that species into which the hybrids merge, but the effect must be only temporary, and no change in the type of the species will result.

We cannot but conclude, that in different cases the course of evolution must be a different one. Still, it looks as if in the majority of instances species are pure for their type, and do not change by continued natural selection, although this selection may have had a marked influence when these species originated. Occasional crossings (between such species as are called sub-species) heighten the total variability of species, and this variability is continually reduced automatically. Wherever it exists, even if it has no influence on the type of the species itself, it makes it possible for a split-off group, an isolated small colony to have a total potential variability different from that of the old species, and therefore, under the influence of natural selection automatically to become pure for its own specific type, its own genotype.

Colonization must be a common phenomenen, and it seems, as if the directive action of natural selection on species were restricted to the very short period during which an islolated group becomes pure, becomes a species.

To become established as a new species, new sub-species, a group of organisms must be isolated, either in space or physiologically, for a time, sufficiently long for the group to become numerically important enough to be able to withstand occasional contact with the species from which it diverged.

If the isolation is a mere isolation in space, a colonization, those new species will stand the best chance to survive, which are as well or better fitted to live than the parent-species. A new species cannot survive if there is not some reason, which makes matings between individuals belonging to it much more frequent than inter-specific matings.

The very fact, that there are common and rare species, testifies against an effect of natural selection upon established species. Those species are common whose genotype makes them fitted to live and procreate. We can roughly distinguish two kinds of rare species, those which are rare, because they are adapted to special conditions which are not often realized, and species which are rare, because they are not well adapted to any environment. Species of the first group are rare locally and may be common in other localities, but species of the second group are rare anywhere. If natural selection within a species were effective, rare species would tend to become commoner, because they would become better fitted to survive.

The inevitable result of the fact that only a fraction of the individuals produced in any species can actually live on earth, is a selection, a natural selection, but if the selected, surviving, procreating group of individuals is genotypically identical with the suppressed group, no change in type will result from the selection.

The essential differences between what happens in cultivation and in nature are soon apparent, if we understand, that for a change in genotype, for an effect of selection, isolation of one sort or another is essential. Whereas in nature, perpetual crossing is the rule, and varieties have only a fitful existence, in cultivation isolation is the rule.

To begin with, the very act of taking a species into cultivation is in itself an act of isolation. The group of individuals which is taken from its natural surroundings and propagated, will have a total potential variability smaller than that of the whole species. I believe it is very seldom that a species can be taken into cultivation and propagated as such, and prove to be a valuable animal or plant. The requirements under cultivation must necessarily be quite different from the requirement in nature. The very qualities which make a plant or animal a success in nature, may count against it in cultivation, and reversely, plants and animals are commonly valued for characters which would debar them effectively from propagating without the help of man. Species which are rare in nature, may happen to make more valuable cultivated plants or animals, than species which have proved to be a success in nature.

As to the effectiveness of isolation under cultivation, it must be remembered that no strict intentional isolation need be practised in a group of individuals, which nevertheless, by the very fact of being cultivated may be sufficiently·isolated to constitute a new species.

If the natives of some uncultivated country catch a number of young wolves or jackals, and make them live in their midst, they take no pains to prevent their mating with their wild relatives. But, notwitstanding the fact that occasional matings of tame wolves and wild ones occur, the number of instances in which tame animals mate with tame is sufficiently larger than the number of instances of crosses between wild and tame, from the very fact that the tame ones live in close proximity to man, to make a species out of the tame group.

If one takes even a common wild-plant like Ray-grass or Thyme into cultivation, the very massing of the individuals in his fields constitutes an isolation, sufficiently close to insure that the group will become pure for its own genotype, as its total potential variability must have been a fraction of that of the whole species. Crosses with wild-growing plants will occur, but if we contrast the close proximity in which the cultivated plants grow to the scattered stand of the wild plants, we see that the very massing is an effective means of isolation.

Selection cannot influence a group unless this presents some potential variability, and unless it is isolated from the multitude. In cultivation both conditions are met.

Why should the potential variation in cultivated plants and animals be high, and therefore make them amenable to change by selection? We saw, that the very fact of their being taken into cultivation made it probable, that the group has a potential variability smaller than that of the whole species. The answer to the apparent paradox is that species in which the potential variability is low, have no good chance of being a success as cultivated animals or plants. For it is significant to note, that it is not those animals or plants which promise the greatest return per unit of area, which have attained the greatest successes under cultivation, but it is those groups with the largest variability, those groups in which some forms are adapted to some, and others to very different uses, or different climates.

Perhaps the goose and the turkey can be kept more economically than domestic chickens, yet chickens are more of a success than either geese or turkeys as domestic animals, and it may be that the oil-palm gives a greater return per acre than corn, yet corn is a greater success as a cultivated plant.

Where, as in nature, the variable off-spring of an occasional cross goes under into the multitude, under cultivation aberrant individuals are apt to be noticed and given a chance to show their value.

Under cultivation both processes in evolution, on one hand heightening of variability by crossing, and on the other hand reduction of variability by isolation, selection and colonization are exaggerated far beyond anything we can ever hope to find in nature.

Propagation of plants and animals under domestication is essentially different from propagation in nature, as the former is essentially a continued system of colonization.

The history of the domestic breeds is one of repeated colonization, of isolation of small groups. It is very instructive to note how a breed of dogs is introduced into a new country, speedily becomes popular and is propagated, and to note from how few individuals the multitude of animals of a new popular breed is derived.

The result is, that such a breed in its new home is very much purer than in the country where it originated, and this very fact may be part of the reason for its popularity. The Airedale terrier in America, is purer and different and better as a breed, than the same breed in Scotland. The Schipperke in America and England can hardly be compared with the original dogs, asthey were kept by barge-men in Holland and Belgium.

Evolution in plants and animals under cultivation is certainly more intense and often more accelerated than evolution in nature. Whereas the origin and establishing of a new species in nature must by the very nature of the conditons which make it possible be a rather rare process, though not necessarily a slow one when it occurs, new cultivated plants and animals originate frequently.

The main point of difference is this, that we have good reason to assume that whereas species in nature may be replaced by others, but do not change under natural selection after they are once formed, domestic species do change by selection.

In most domestic animals and plants there is no pure, unselected multitude into which varieties merge and which constitutes the type of the species. Propagation under the favorable conditions of cultivation is so quick, that the whole mass of individuals of a certain strain is commonly descended from very few individuals a few generations back Under the hands of a few breeders every sub-breed becomes a species and it changes rapidly. A few, a very few individuals who are more like the pre-conceived ideal toward which the group is bred, are carefully bred, and it is seen to, that their progeny is as numerous as possible so that the whole group varies in their direction. In most domestic animals only a very small minority of the males are used for breeding at all, and there selection of a few males has a very great influence on the whole breed. In all the cultivated animals and plants there is always enough cross-breeding to keep up the potential variability necessary for a further change under selection. Even in the more highly-bred animals occasional animals are registered, and therefore taken up into the groups which here are species, which have a more or less remote ancestor belonging to some other species.

In this connection it becomes necessary to reëxamine the proofs which Darwin adduced for the monophyletic origin of some of our most variable domestic animals. For it stands to reason that, if it is true, as it appeared to Darwin, that selection within one species can produce such different animals as a Jacobin and a Fantail pigeon, or as a Silky and a Polish fowl, selection in nature must influence species and modify them continually. Darwin conceded the polyphyletic origin of the

domestic dogs, because wild Canidae have been repeatedly domesticated and because hybrids between dogs and different
wild canidae have proved fertile.

The groups of domestic animals for which Darwin assumed a
monophyletic origin are the fowl, the pigeon, the rabbit and
the canary. The reasons which led Darwin to a belief in a monophyletic origin of all the domestic breeds of tame fowls were
the following. No wild fowl now exists which exhibits the very
marked peculiarities of very many tame breeds, such as the
Dorking's fifth-toe, the feathered-crest of the Polish, the frizz-
led feathers of some Japanese bantams.

All the tame chickens are mutually fertile, they all have approximately the same voice and in most breeds there are sub-
breeds coloured like the wild Gallus bankiva.

Is it possible, in the light of the experimental evidence which
has become known since Darwin, to maintain the possibility
that all the tame fowls descend from one wild species? I think
not. No wild species by itself has a potential variability sufficient to account for the variability in the tame chickens. Let us
examine Darwin's ground for a belief in the monophyletic origin of the tame breeds in detail.

In the first place it is undoubtedly true, that no wild chickens with five toes, or upturned feathers, exist. We must not forget that Darwin's thesis, that all tame chickens originated from
one species was originally meant to refute the idea of the fanciers that, every tame breed descended from a separate, now
extinct, wild species. Even comparatively recently it was
urged by Davenport that the Malay breed must have originated
from a species, different from that from which all the other
chickens descended. As far as it goes, Darwin's view is the more
logical one.

But, and this to our mind is the most important new fact
bearing upon the problem, we now know that cross-breeding
produces absolutely new characters. Just as we saw double-
flowers and fimbriation, and polycephaly result from a cross
between two species of Argemone, and just as we saw waltzing

originate from a cross between two species of rats, we may expect very striking new characters to result from a cross between two wild species of Gallus. If, therefore, the hybrids between tame-chickens and a species of wild-chicken, other than Bankiva, are fertile when crossed back into the tame stock, this fact is sufficient to account for the remarkable variability in the domestic fowl. Darwin adduces an instance of a hybrid G. Sonnerati-bankiva which was not completely sterile. We know now that partial sterility is no bar to reproduction, and eventual full fertility of the descendants. This is clearly shown by the experiments of Bellings with Lyon beans, and of Detleffsen with guinea-pigs.

During our stay on Java we observed numerous hybrids between Gallus varius and tame fowls. These hybrids are produced in regions where the natives make a speciality of them. On some of the islands Karimon Djawa, laying off the North coast of Java, such hybrids are produced in great numbers. They are commonly produced by pegging a tame hen down on the ground near a bottomless basket under which a tame Varius male is kept. A tame Varius hen is introduced under the cage, and when the male mounts this hen to copulate, it is withdrawn and the cage with the male pushed over the tame hen.

The hybrids are called "Bekisar" and they are very highly prized both by the natives and by Chinese and Arabs. They are commonly kept in bamboo cages slung high in the air on bamboo poles, where they waken early and crow. Very high prices are sometimes paid for individual birds with particularly attractive voices. These Bekisars are by no means sterile. In Pasoeroean, where a great number of Bekisars are kept by Chinese, we saw several very beautiful cocks in cages, bred from Bekisar fathers. We remember a white male with a long bluish tail and a blue-bronze neck, of which every feather was rounded as in Varius and bordered with black, and an enormous black cock kept by a Chinese carpenter, that had the blue comb and median wattle of Varius.

In Ketanggoengan West we saw two broods of chicks, whose father was a reddish Bekisar with bluish wings, median wattle and unserrated comb. This Bekisar (there were two males) was bred from a buff-bantam male and a female Gallus varius.

Charcoal-burners often take a few hens with them into the jungle, where they mate mith Varius males, and produce Bekisars. They claim that hen Bekisars, which have no value commercially, are fertile with wild males, and a certain number of the apparently pure Gallus varius offered for sale on Java are assuredly produced by a sort of "Grading" process, a repeated back-crossing to Varius males.

If questioned as to the origin of the tame chickens, the natives of Java will declare Gallus varius to be the wild progenitor of all tame breeds. Although this is undoubtedly untrue, it is certain that a great many characters common to Varius are quite common in the Kampoong chickens on that island, such as a single median wattle, unserrated comb, blue and yellow tinge of the comb, round hackles It would be interesting to find out just how much variability would result from a cross with Varius. It is certainly worth noting that on Java, where hybrids between the two species are continually taken up into the population of domestic chickens the variability of these animals is stupendous. Not only do we see all shapes cf comb, and all colours common to chickens, although the colour of the wild Gallus bankiva is rather uncommon, but we saw characters in Java which we never noticed anywhere else. To enumerate a few of these: A common sight is a hen with only the feathers of the neck turned up, and we often saw animals in which the feathers of the back were also recurved. Chickens with more or less complete loss of feathers are commonly met with, ranging from animals with bare necks like the Siebenburger breed, to fowls that are completely naked with the exception of a dozen feathers on each shoulder. The penguin-like carriage of the Pouter pigeon and of the Runnerduck is sometimes seen in fowls. In such animals the head is kept well back of the centre of gravity, and they walk with out-spread legs in a peculiar shuffling way. Another remarkable variety, which is not rarely seen in the East of Java, is represented by animals in which all the feathers are reduced to bare shafts. It is certainly remarkable that all these aberrations, together with the better known ones, such as absence of tail, drooping-tail, foot-feathering, complete frizzling, occur where crossing with Varius takes place. We would not be understood to say, that we believe that the tame chickens descend from hybrids between bankiva and varius, from Bekisars, and we would not hesitate to call the tame fowl Gallus bankiva hybrida. As we see the facts, we would say that the tame fowls are descended from domesticated Gallus bankiva, the potential variability of which was, and is still heightened by taking up of hybrids with *varius* and not impossibly with other wild fowls into the species.

The second domestic animal for which Darwin assumed monophyletic origin is the pigeon. We have to concede Darwin that all tame pigeons have several characters in common with Columba livia and that no other wild pigeon exists which would be more likely to be the progenitor of all tame breeds. Here, as in the fowl, we need not look for wild species showing the various characters of domestic breeds, but, as in the fowl, we may assume that the necessary variability in pigeons was, and probably still is, produced by cross-breeding with other species, if we can find instances of hybrids which produce fertile offspring when mated to domestic pigeons. We do not need to restrict our search to the species which nest in cavities, as Darwin believed. For it is clear, that the descendants of hybrids with a species nesting in trees, which would have a genotype compelling them to nest in trees, would automatically get weeded out of the population. Mr. Podmore has shown that hybrids between the European Woodpigeon, Columba oenas, and tame pigeons can be bred back to tame pigeons. Therefore, we can at least look to this species as to one of those, that can have given the great variability in domestic pigeons. It may look strange to assume that the ultimate origin of such aberrations as that of the Fantail, or the Pouter or the Maltese may be sought in a cross with a Woodpigeon, but if we see how walt-zing rats originate in the second generation of a cross between two outwardly identical species, the Javanese and the Suma-tran field-rat, this hypothesis loses much of its improbabliity.

We cannot continue to believe that all the tame breeds of pigeons are derived from one species. As we did in the case of the fowl, we think we can here quite logically call the domestic pigeon by the name of Columba livia hybrida, and state that the total potential variation of Columbia livia has been heightened by crossing with Oenas and possibly other species.

The third case is that of the duck. So many cases of fertile duck hybrids are known, that it is really quite superfluous to believe that all tame ducks must have descended from Anas boschas. As we know that absolutely new characters sometimes originate as thè result of crossing, there is no great difficulty in the way of the assumption, that the variability of Anas Boschas, after it was taken into cultivation, was repeatedly heightened by crossing with other species, Dafila acuta for instance. We now know, that in such a case we need not look for characters belonging to these other species in our tame ducks. They may be absent. If after a cross with Dafila acuta animals with a top-knot are produced, or albinos, or blacks, or penguin-like animals, such animals have assuredly later on, been crossed back into the old species, and only the aberrant, varietal character is retained.

The case of the rabbit is somewhat more difficult, because the only species which can be drawn into account in Europe to have heightened the variability of the wild rabbit, is the Euro-pean hare, Lepus timidus.

Experiments have been tried in several Zoölogical gardens to mate the hare with rabbits, by accustoming the animals to-gether when young, but hybrids have never resulted from these trials. In the first place, we must not lose sight of the exist nce of very many different species of wild rabbits, in Europe as well as in America. The assumption, that from a cross between the

European cotton-tail with a similar American species, animals of a black colour, or giants, or rabbits with extremely long ears, should have resulted, is not so very out of the common, if we see how yellow and chocolate and silver rats, and waltzers originate from a cross between two or threee sub-species of Mus rattus.

As to the cross between tame rabbits and the European hare, we know that Belgian and French rabbit-fanciers are convinced that the cross is possible and that it has been made in both countries. There are tame rabbits which very closely resemble hares, and which certainly look as if they had descended from some hare ancestor. We have tried the cross but failed, the difficulty lying, we think, in the fact that a male hare will not easily breed in an enclosure, being too nervous.

In France it was told us, that a male hare in an enclosure will mate with female rabbits, provided the hare is given a dark house in which to hide, and the person who introduces the rabbit hides himselt. We feel certain, that it is only a matter of technique to produce these hybrids.

The wild rabbit, such as it exists in a state of nature, is in no way a desirable domestic animal. It is next to impossible to tame one, and we do not know of a single instance of their breeding in hutches For this reason it seems probable, that only after some cross, the animals have become sufficiently variable to admit of domestication The tameness of animals is certainly not a matter of domestication. We mean, that no long series of generations in captivity will make any wild animal more tractable, if it was pure to start with. Rats of the species Mus concolor, and Sumatran field-rats are just as wild and untractable after several generations of captivity as the wild-caught animals. But in the house-rat group, where we have crossed several species, we have by a sort of unconscious selection of those animals which would breed in comparatively small cages, obtained a strain of animals which will breed quite readily in captivity and which can easily be made tame We doubt very much whether any wild animal would make a satisfactory domestic one without crossing of some kind at the outset. As in the case of the fowl and the pigeon, we do not need to look for wild rabbits or hares with the markings of the "Papillon", or with ears that measure 24 inches, but we may rest assured that crossing with even a very similar American cotton-tail, might produce such aberrant characters.

The thesis, that no extensive variability, and therefore no formation of very many different domestic species is possible within one good single species, is further strenghthened by comparing the variability in such poly-genetic groups as the swine, cattle, dogs, to those animals that have been long domesticated, but which are almost certainly descended from one wild species, such as the guinea-fowl and the pea cock. We know two recessive colours in the guinea-fowl and three in the peacock, and no structural variations whatever, and we may with some reason assume that these few variations are the result of loss-mutations.

Contrary to the belief that the semi-domesticated state in which guinea-fowl and pea-fowl have been kept, hindered them from varying, as they thus were not subjected so much to changed conditions and different food, we are convinced that these animals are only semi-domesticated *because* they did not vary through crossing. If they had been crossed, each with a related species, they would not only have varied in shape and colour, but also in their adaptability to life under intensive cultivation, and selection would speedily have made them domestic animals.

There are great differences between the course of evolution in domestic plants and animals, and in species under nature. In nature, every species consists of a multitude of like individuals, and matings are greatly a matter of chance. This results in a great stability, and it makes selection of any kind ineffective within a species. Natural selection, plays its rôle evidently at the origin of new species, that is to say, that natural selection will evidently help to determine the type, for which an isolated group of individuals, with a sufficiently large total potential variabliity will become pure. We picture the origin of species in nature as brought about by any cause, isolating a group of organisms from the swamping effect of free-crossing with the species multitude, provided the total potential variability of the isolated group permits of a new genotype, a new combination of genes.

It is evident that any small group of individuals as such, must have a potential variability smaller than that of the whole species at the moment of isolation. Therefore it may seem paradoxical to assume that under the influence of selection and automatically, such a small isolated group can give rise to a new species, whereas the old species remains unaltered.

But it must be remembered that in such a small isolated group, if it has any potential variability, that is to say, if it is isolated at a moment when the total variability of the old species has recently been heightened by a cross, the individuals of the common specific type do not constitute a great preponderant multitude.

Whereas species in nature are stable, and do not change by natural selection, it is easy to see how they can gradually be replaced by new species, springing from them, reëntering their territory and which prove to be better fitted to live in the old conditions of life. From paleontological evidence it is impossible to decide whether a species has changed as a whole, or whether species have repeatedly crowded out their parent-species.

In cultivation, the main difference from what happens in nature, is given by the control of propagation. There is no multitude of geno-typically identical individuals into which variants merge. Variants of some merit are selected, and continually a sort of colonization is taking place, starting with very few individuals of a typical constitution. In cultivated animals and plants the species can change as a whole. An extreme, but very illustrative example is furnished by cattle. Here aberrant, from the breeder's standpoint superior animals are not only selected and mated to similar aberrant individuals, but very often the exact reverse takes place from what happens in nature. We mean, that not only are these aberrant individuals hindered from being taken up into the multitude of average individuals by repeated crosses, but on the contrary, very many average, typical individuals are taken up into the new species. They are mated to the few very good individuals, and their children also and so on. The change of species under domestication is not only one by colonization, but actively a change of the whole species under influence of selection.

It is evident, that we must be very cautious in concluding as to the course of evolution in nature from facts observed in cultivated animals and plants. Certainly the two processes are not identical, and from the tact that under cultivation species are changed by selection we may not conclude that they are similarly changed by natural selection in nature. But if we observe the cause of the difference, the existence in nature of an inert, unchangeable body of individuals in every species, and the absence of such a multitude in cultivation, we have further proof for our contention that species are on the whole stable, unchangeable, but that new species differing in small, often adaptative changes from parent-species, are continually being formed in those cases where crossing furnishes the necessary potential variability, and isolation furnishes the chance for the new group of settling down to its own genotype.

New species can become established only if three conditions are fulfilled. The potential variability of the parent-species must be high enough, the new group must be sufficiently isolated from inter-breeding with the parent-species, or closely related species, and thirdly, the genotype for which the new group becomes pure must be such as to insure its members a reasonable chance of surviving and procreating.

If we consider the possibilities and probabilities of species-formation it becomes apparent, that different groups of organisms must have different chances to produce new species, that the course of evolution is not the same in all groups of organisms. Rate of reproduction, rate of dispersal, colonizing hab-

16

its, all influence the possibility of evolution, but it is evident that the most important division of all organisms in respect to their chances of evolution is that between organisms which cross, and organisms which do not habitually cross.

In organisms of the latter category, any individual is a potential species. If its genotype insures it a good chance of life, its descendants are absolutely isolated from other lines of organisms. One of the three conditions for the origin of new species is therefore in these organisms fulfilled under any circumstance and in any environment. Another fundamental difference between organisms of this group and crossing organisms is, that the potential variability of every species is very soon zero. Therefore one species cannot have daughter-species which are directly and exclusively derived from it. The potential variability which are directly and exclusively derived from it. The potential variability necessary for the production of species with new genotype is possessed only by hybrids. In other words, species in this group necessarily have a poly-genetic origin, whereas in allogamous animals and plants at least the possibility exists of the origin of a new species from an original one.

Groups of organisms which are quite closely related, may yet differ in this respect, so that the mode of evolution in the one differs fundamentally from that in the other group. Wheat for instance, is almost exclusively autogamous, and new species in this plant can arise only from hybridizations between species. On the other hand in this group, almost any individual plant is a potential species. Rye crosses freely, and in respect to evolution it is fundamentally different from wheat. The origin of new species in rye is closely similar to that in animals, and yet, wheat and rye are similar enough to admit of crosses between them.

In the evolution of species of those organisms which do not habitually cross, isolation is provided for, so that, whenever there is a heightening of the potential variability as the result of a cross, numerous new species can originate. In the establishment of those, the actual fitness of their genotype decides the possibility of their survival. New species in these organisms can establish themselves in the territory of their parent species, living in identically the same circumstances, fitting in the same ecological niche.

The biological law proposed by the zoölogist Jordan, that closely related species will be found either in different territory, or fitting different ecological niches in the same territory holds good only for allogamous organisms. In making his generalization he did not consider such species as we know in autogamous plants, and as were first demonstrated by the botanist Jordan.

Species formation in allogamous organisms is a process of colonization. Evolution, the establishment of new species in allogamous organisms is practically impossible within the same territory and the same ecological niche already occupied by a species. When a certain number of individuals of a species colonize in a place where conditions are right for them, the barriers separating the new group from the multitude of the species may be strong enough, and (or) the rate of disperal of the organisms may be slow enough, to make matings within the group far more common than matings with individuals of the old species. If this is the case, the possibility exists for the origin of a new local species. The potential variability of the new group will reduce itself, and the final type may be the original type of the species, in which case we again have a case of colonization. But the final type may also be somewhat different, if the potential variability allows of this. The new group may have a slightly different size or colour or shape, or rate of reproduction. Natural selection may have had nothing to do with the final type of the new species. The new characters need not be of any advantage or adapt the organism any better to its surroundings. In a study of the smaller mammals, and of birds, such local species which differ in trivial characters, and occur in the same ecological position but in different regions have been found to be the rule. The American zoölogists have collected valuable data on this point.

The isolation required to make the proportion of intra-specific matings to inter-specific ones sufficiently high to make a species of a number of individuals is brought about by a good many factors, which sustain and counteract each others action. Actual geographic barriers are not necessary to keep species apart. Organisms with a quick rate of reproduction and with a quick rate of dispersal will probably require actual barriers to keep species apart. But it is evident that a slow moving group of organisms may be differentiated into several species without the existence of geopgraphic barriers of any kind.

When we speak of isolation as of the necessary requirement for species-formation we do not quite follow Wagner in assuming the need of an actual isolation in space. Isolation simply, is the most common cause for the required proportion between the number of intra and inter-specific matings. But it is evident, that any other cause bringing about this proportion acts in this way.

In a slowly dispersing animal the chances for individuals wandering out of the territory of their species into that of a neighbouring one is small. Snails are for this reason especially apt to form local species, very much more so than weasels.

New species can only originate in the territory occupied by a parent-species, or by a closely related one, or they can only reënter this territory if they are for some reason protected from crossing freely with it. By far the most common case will be the one, in which a new species fits into a somewhat different ecological niche, but by no means the only one. A difference in size may preclude crossing to an extent sufficient specific distinctness. Or a difference in mating season, or one in the structure of the sexual organs may have this influence. The case of the two house-rats differing in size in Java and British India, is a good instance of two distinct species, very closely related, fitting the same ecological niche and yet remaining distinct as species.

If we should state with Jordan that two very closely related species can only occur separated by a barrier, or if we enlarge the statement by adding — or fitting different ecological environments, we would not state the whole truth. We would certainly have enumerated the main causes for specific distinctness. To go to the root of the matter we have to state, that species remain distinct, when the automatic reduction of the potential variability of the groups through any or all ·reasons, outweighs the heightening of the potential variability by crosses or otherwise.

Evolution, the establishing of groups with new genotype is impossible within freely crossing populations. Whenever there exist causes which limit free crossing to a certain extent, and which depend upon nature of the factors reducing the potential variability, such as colonizing habits, rapid reproduction, evolution can proceed. Geographic barriers, slow dispersal, autogamy, different size etc. are all at certain times causes for the differentiation of species. And we have seen that neither of these causes is in itself indispensible, one can replace the other in the establishment of the final balance between the factors heightening and those reducing the potential variability of groups.

Is it possible from an inspection of groups of organisms as they occur in nature, to estimate what factors from among the list of those possible have coöperated to establish them? To a certain extent it is.

If we find sharply demarcated groups of closely-related organisms differing in trivial characters, such as the local species of deer-mice or ground-squirrels, we may feel safe in assuming that just chance determined for what type, possible of realization in the potential variability of isolated colonies the group has become pure, and if intermediate forms are absent we can feel reasonably sure of the existence of real barriers.

Should we on the other hand, meet with forms which grade more or less gradually from one local species to the other, the existence of barriers becomes doubtful, and we have reason to suspect that the animals in question simply keep from mixing because of a slow rate of dispersal. In such cases where barriers do not exist and where local species gradually merge one in to the other, we meet a state of things different from that in which local species exist on both sides of a region not suitable for occupation for these organisms, or where local species fit different ecological niches in the same locality. In the latter cases the species differ in groups of genes, whereas in the absence of barriers the gradual change of type which we observe in passing over one continuous region, is due to the fact that differentiating genes are present in groups of organisms, which groups do not correspond for the different genes. Each differentiating gene, by which we mean each gene which is present only in part of the organisms of the whole group, has a certain territory. Whenever we are dealing with a gene, which has such a marked influence upon the characters of organisms carrying it, that the difference caused by its presence or absence is greater than any chance difference through environmental influences, we note discontinuous variation in some character in addition to continuous variation in a number of other characters. But where we are dealing with genes whose influence is less, or with several genes which influence the same character in the same or in opposite ways, the differences produced by their presence and absence are gradual.

It is extremely difficult to decide which groups in such series of organisms should be called species. The safest way, we think, is to give specific names to as many types as can be distinguished. These species with a limited range and no very sharp boundaries are certainly not varieties.

Whenever we find two or more species of obviously very closely related species living in the same territory, we know that something must interfere with their free inter-crossing. If autogamy is excluded we will find that in most cases the two species live a somewhat different life, they live in the same geographic region, but in different spots, they have habits which bring members of the same species into very much closer contact than members of different species. But sometimes we find other reasons for the high proportion of intra-specific matings as compared with inter-specific ones. The two species may differ in size, or they may react in a different way to the seasons so that their mating periods do not coïncide.

In attempts to hybridize species, a difference becomes apparent between the various sets of species found in nature.

When two species coëxist in one environment, and when they are found to live the same life, and to occur in the same spots, hybridization will for some reason be practically impossible. It is apparent that we may look for sterile hybrids in such cases, though we have not met which such an instance. Very often in the case of animals, no matings occur. The case of the two house-rats of Java and the Malay peninsula, Mus griseiventer and concolor is a typical instance.

Where local species of closely contingent regions are crossed, they will mate as readily as members of the same group. And the difference between them will be found to be comparatively slight, comprising a few genes. The cross-breeding experiments of Sumner with Peromyscus have shown that the genotypic difference between local forms is slight.

Lang's experiments with snails, Helix, prove the same thing.

When two closely related species which inhabit the same territory, but living a different life are tried, it will sometimes be found to be almost impossible to cross them. Such a case is that of the Javanese field-rat and house-rat. In other cases, such as that of the house-rat and tree-rat, hybrids are easily produced from caged animals, which shows that the reason for the specific distinctness must here be ascribed to the different mode of life. In the case of the house-rat and the tree-rat, the result of the cross shows that the genotype of the two species is more different than between contingent local species. In the second generation unexpected new characters are seen in several animals, just as in the case of a cross between two rather widely different species of plants.

This difference in genotype between closely related species occurring in different ecological niches in the same environment must be the result of the fact, that they are effectively isolated one from the other, so that they do not share in the heightening or changing of each others potential variability and each others genotype.

Two groups of animals separated by an effective barrier may be closely alike pheno-typically, and yet, their genotype may be different. This can be seen whenever individuals of two such groups are crossed. We found the field-rats of Sumatra and Java, which are apparently identical to be different geno-typically by crossing them.

THE STATUS OF MAN.

MENDELIAN inheritance in man was demonstrated almost simultaneously with the first work with animals, which showed that the clue discovered by Mendel in his work with plants, was going to be of the very greatest importance for an insight into heredity and evolution.

It was soon apparent, that differences in characters of mankind were often due to uni-factorial differences in genotype, and that such differences could easily be traced through long series of generations. Apart from differences in eye-colour and hair-colour and such, most of the cases of inheritances of definite genes in man have been those in which the lack, or in other instances, the presence of a gene manifested itself as a pathological aberration. A few instances are colour-blindness, brachydactily, alkaptonuria.

Most of these studies on inheritance of characters in man were at first conducted out of pure love of knowledge, without afterthought, but, as some of the characters studied happened to be undesirable ones from a medical or a sociological point of view, the thought lay close, that some knowledge concerning the inheritance of these characters was the first step toward eliminating, or at least combating them.

Of late years a good deal of interest in Eugenics, is evident, and it is safe to say, that almost all the persons interested, hope that from a study of inheritance in man some good for the future of humanity may result. Only a limited number of investigators just happen to study the transmission and interaction of genes in man, instead of in some other organism, but as from a genetic standpoint the material has very material drawbacks, man becomes less and less popular as a subject for genetical investigations.

It is the object of this chapter, briefly to examine the method of the Eugenists, and to discuss what hopes there are, that these methods will bring us nearer to the desired goal. And we will further examine the status of mankind with the aid of the conception of species developed in the preceding chapters, and try whether it is not possible to open-up more promising avenues of research.

All those cases, in which the inheritance of a definite character is followed through an extensive number of families and for several generations, have been concerned with qualities, differing from normal in the lack of one, or in the possession of one more gene. The limitations in the material, the small number of descendants from one pair of individuals, the impossibility of making test-matings, the doubtful reliability of the record in many instances, preclude more complicated cases being succesfully studied.

The expectation has been, that it would be possible gradually to progress from more simple cases to more complicated instances, and so to work out the inheritance of very many inherited qualities in man. These hopes have not been realized. A great many more very simple cases have been added, and the first studied instances have been proved over and over again. When we are dealing with a plant or a small animal of which we can breed large numbers, it is possible to work out the action and inter-action of a great number of genes, and from a close control of our stock we can safely conclude as to the identity of the genotypic peculiarity which results in the same quality in a number of individuals. Ten unrelated pedigrees added together to make two-hundred individuals showing the inheritance of some character, have not the same value as one family of forty individuals. A great number of cases of colour-blindness may be identical from a clinical standpoint, but it remains to be proven that they all depend upon the same genotypic peculiarity.

For this reason, the stupendous amount of work done in coöperation by numerous zealous investigators, in adding and adding data on the inheritance of mostly the same aberrations, satisfies us so little. In itself, as a study of the behaviour of two, or generally of only one gene, these studies are not of as much interest as the inheritance of black and grey colour in as many mice of one strain. The only thing which makes this work interesting enough to attract so many investigators, is the fact that the subjects of the study are human, and that the characters under consideration affect human well-being. This means, that most of these studies are now performed with no ulterior motive but the obvious one, to know more of the transmission, and thus of the possible control of undesirable characters.

The discovery of alternative inheritance in man raised great hopes for the development of Eugenics. In the first place this evidence proved once and for all, that inheritance played an appreciable role in the development of the individuals as it held out the possibility of finding a tangible basis for work on the "improvement of the race".

Much has been written about this "improvement of the race". It has struck a great many authors as illogical, that, whereas man gives so much thought to the improvement through selection of his domestic animals, he should leave the future of his own kind wholly to chance. It is entirely obvious, that the methods which developed the improved dogs and horses would work out in a similar way in man. But from the majority of the writings on Eugenics it is clear, that the authors interested, have no clear conceptions of just how these improved dogs and horses originated. A very readable discussion of the subject, and a critiscism of the impatience of the Eugenists is contained in H. G. Well's book on Mankind.

Selection of the best individuals and elimination of the least good must result in an improvement. This seems evident. At first sight it appears, as if any encouragement of the propagation of the best individuals, and any elimination of the less desirable must help toward improvement. But we have seen,

that those forces, which produce the automatic reduction of
the Potential variability of a group, will tend to nullify the
influence of a minority. Unless therefore, the proportion of
undesirable individuals becomes greater than a certain minimum, these individuals will eventually have no influence upon
the type of the group. And on the other hand, occasional superior individuals have no future, and unless their proportion
to the total number can be brought up vey considerably, no
good will result from encouraging them.

As soon as we begin to make a comparison with animals the
question arises, do the breeders work by favouring the reproduction of the best animals and by discouraging the breeding
of the less desirable ones? A little later we will discuss the
question, whether this is the procedure in the development of
each breed. But first, can it be truly said that, let us say horses,
have been developed by this method, from which so many
authors await improvement in mankind?

If the breeders of animals work in this vague way, breeding
from the best and as little as possible from the less good, how
does it happen that so many different breeds of horses, of dogs,
of poultry exist? It would appear that, if this selection by man
of the best dogs were consistently followed, eventually all the
dogs, or at least all the dogs of one region would approach to
the ideal type. To take a concrete example. Breeders of Spitz-
dogs abhor cream-coloured animals. Is this a general rule in
dogs, that cream colour is undesirable? Evidently it is not, or
there would not exist breeds of dogs which are pure for just
this colour, which is so frowned upon in the Spitz.

To any animal-breeder the idea that anyone should conceive
of an ideal dog, an ideal horse, and would propose to breed the
most from individuals nearest to this ideal, is inconceivable.
There is no ideal dog. There is however, an ideal Coach-dog,
or an ideal Grey-hound. There is an ideal Hunter and an ideal
Shetland pony, but there is no ideal horse. The whole group,
all those animals which are counted as horses in an agricultural
census, consists ot several distinct breeds, plus some hybridized

individuals. We have seen, that these breeds are species in every essential sense. And the dogs of one country fall into a number of species plus a variable number of animals outside these species, hybrids. And if we consider selection, we see that the groups which are subjected to selection by the breeders are these species, and not the compound groups, not the whole horse-population, or all the dogs of a country.

Now it is to be deplored, that most of the studies by genetocists on domestic animals have been on the inheritance of definite single points rather than on the causes for species-formation, or on the improvement of species by selection. In other words, most of the work has been done with varietal characters. It is obvious, even if we know in one species of dogs, how blue, black, cream and yellow colours are related one to the other, there is no connection between such knowledge and what we would have to know to enable us to improve the dog-population of one country as a whole.

All work on the improvement of dogs or poultry logically starts with a study of the different breeds, the species. And the man who would try to improve the dogs of a country as a whole, would be laughed to scorn by the dog-fanciers. It would be equally impossible to lay down rules for the improvement, not of one or the other species of fowls, but of chickens in general. And yet, people who propose to find methods and regulations to improving the whole human population of a country, or even to improve all mankind are listened to with respect.

There would be one exception to the rule, that it is impossible to improve the whole dog-population, or the whole horse-population of a given territory as such, namely in the case in which there was no differentiation into species, specialized breeds, and if such specialization were not desirable. In such a case it would be possible to establish a set of ideals, and breed towards it. The reindeer in Alaska may be bred in this way.

And it is evident, that even if some authors would not go so far as to believe in the specific unity of all mankind, the specific unity of one people is very generally taken for granted. With what right we will later discuss.

The very first move in the improvement of the live-stock of a country is a study of the different species, the uses to which they are put, the possibility of making them fit these uses better. In certain cases amalgamation of two or more species may be seriously considered. If our object is the amelioration of the dog-population, or of all the horses, all the fowls, we have to deal with the typical qualities of the species, and such details as the relation between sorrel and chestnut colour in one breed of horses, or the inheritance of broodiness in one group of fowls have practically no interest.

To one who does not know dogs, the variability of dogs seems kaleidoscopic. All combinations of the most diverse characters are seemingly observed in a collection of dogs by anyone not conversant with the characters of the different species. A Zoölogist at a dog-show will see dogs with long pointed heads, and others with very short jaws, some with little twisted pigtails and others with long bushy tails. It is a very curious experiment to take a Zoölogist, who does not know domestic animals, to a good dog-show or poultry-show, and afterwards to make him talk about what he saw, an experiment which we can recommend from experience. The thing which is apparently most interesting to the trained observer of animals, is the enormous variability. The stupendous diversity will strike a Zoölogist, but, curiously enough, the existence of relatively little variable groups within all, this mass of animals will on the whole escape him. If he attempts to describe from memory, what he saw at his first visit to the dog-show or the poultry exhibition, he will remark upon the enormous range of variability in colour and shape, of size in dogs, of tails in dogs and pigeons, of feathering in poultry, of shape of comb, and feet, and beaks. But he will very likely do so in terms, which will to one who stayed at home, give the impression that almost all possible combinations of these characters were actually to be seen at the show. He may not actually have seen a dog with

a little twisted tail like a pig and long hanging ears, or a black
and white spotted Chow, or a silver-spangled Leghorn, or a Fantail-pigeon with feathered legs, but he will certainly expect to
see such combinations at the next show. Somebody who knows
dogs or poultry, will always classify a number of animals into
members of a few species, plus a number of animals of apparent
hybrid origin. The variability in a given group of dogs is not
actually smaller than it appears to be to the Zoölogist, who
does not know the species, but there is vastly more order in it
than he would suppose.

The difference between a red and a black Cocker-spaniel may
appear to the visiting Zoölogist to be of the same order as that
between the red colour of a Cocker, the black and tan of a
spaniel, but to the dog-fancier it is quite another difference.

It is quite impossible to understand the situation and rightly
to valuate the variability of the domestic dog, of the domestic
poultry, unless one first understands the fact, that numerous
different species exist. And it is impossible to understand the
workings of selection in the amelioration of domestic animals
unless it is realized, that it is these species which are improved, and not higher units, such as populations of dogs, or all the
poultry of one country.

The selection of domestic animals has served several authors
as a model in speculations upon the improvement of mankind
by encouraging desirable parents and restraining undesirable
ones.

There is no doubt whatever, that in a group of animals with
a sufficiently high variability, selection in the way in which it
is practiced here, that is, by allowing only the occasional animal of exceptional merit, or at least only the occasional male,
to participate in reproduction, will bring about a speedy change
in the desired direction.

But one or two things have to be remembered. This improvement of a domestic species by selection can be copied from
one variable species to another, but a combination of species,
a group of higher order can not be improved by selection in

this way. If we start with a mixture of two or more species and subject it to selection, it is possible to conceive, how through selection, eventually only one species will survive. But in some cases the inter-relation between the species may interfere, selection in such a mixture is not such a simple matter as in one species, one breed of animals.

When two species occur intermingled, there are forces at work which differentiated these two species, or at least which make them continue to be separate species, and these forces will certainly interfere with the result of a selection to which the mixture is subjected.

If we are dealing with a mixture of two species, which differ in average size, selection for greater size will result in a greater proportional increase of the larger species over the smaller one, but this result is quite a different thing from a change in size of either of the component species. Statistically, one thing is like the other, though biologically there is a vast difference.

In almost all the speculations about selection in human-kind, for so far as they are not concerned exclusively with the inheritance of details, it is casually assumed that mankind, or at least that any considerable section of humanity considered as one whole, is one species. No special pains are taken to prove this point, its importance must be rated very low.

If we examine mankind as a whole from the standpoint of the systematic Zoölogist, what should be the rank accorded to the group, in other words, to what group of organisms should this group be considered equivalent? An objective standpoint is difficult to attain. Just because we are men, we are very apt to divide animate nature into plants, animals and man, for everyday purposes. This, of course, from a zoölogic view, is pure conceit; Rolf, the famous speaking dog of Mannheim, divides animate nature into plants, animals and dogs in just this same spirit.

In his first edition, Linnaeus places man in his system by the side of the Orang-outan, as two species of equal rank.

These are the questions with which the Systematicians should be concerned: Is man one species, Homo sapiens, with a number of varieties, or are we concerned with a number, say four or eight species of men, Caucasian, Polynesian, etc, or, have we a very great number of less extensive species of approximately the same systematic rank? Finally, we may ask whether possibly the situation in man is so different from what exists in plants and animals, that we cannot speak of species in the way in which we use the terms in other organisms.

Is mankind considered as a whole, a group equivalent in rank to let us say the brown rat, Mus norvegicus, a species with some varieties, or to a group like the carnivorae, with numerous species grouped in complexes of a higher order, or to a group of animals like the domestic cattle or the dog, the horse?

Are the different types of man species, or varieties?

What is a variety? A variety is a number of individuals which differ from the type of the species to which they belong in the same way, without a necessary genealogical continuity. There exists a typical species, a ground-squirrel, and all those squirrels together which are found at rare intervals among typical ones, and which are of yellow colour, constitute a yellow variety of this squirrel.

Are the breeds of domestic dogs varieties? There is no such thing as a typical dog, with aberrant individuals which we can group in varieties. There is, however, such a thing as a typical Airedale-terrier, and an occasional blue individual in this breed represents a blue variety of the Airedale-terrier.

Are the groups of man varieties? There is no such thing as a typical, standard man, with aberrant individuals in varieties. There is, however, a typical Zulu-kaffir, and an occasional albino Zulu represents an albino variety of Kaffir; a typical Sicilian, and an occasional feeble-minded Sicilian represents a variety. According to this reasoning the Airedale-terrier, the Zulu, the Sicilian, are species, rather than varieties. Can we group the types of man into half a dozen species, or can we group the dogs in this way? We may conceive of an ideal, a typical Mongolian and a typical Malay. How about the Japanese, is the population of Japan a variety of the Mongolian species or a variety of the Malay species? In dogs we may conceive of a typical hound and a typical terrier, two species. Is the Dachshund a variety of the hound or a variety of the terrier-species? Does not this classification of domestic dogs exist simply to admit of dividing a book on dogs in chapters in a convenient way? Is not the same true in mankind?

What keeps species different? All those causes which make the number of matings between members of one species sufficiently more frequent than matings between members of different species. And it obviously depends upon the factors which reduce the potential variability of a group, how much inter-crossing is constisent with specific diversity.

To take a concrete instance, why are the Airedale-terrier and the Grey-hound separate species? What keeps them separate? We know that hybrids are perfectly fertile, and that there is no preferential mating in these dogs. We find that most dog-owners keep males, and that the females are mostly owned by breeders. These owners of the females guard the mating of their animals, so that almost every puppy raised from either an Airedale or a Grey-hound mother is sired by a male of the same species. There is no inter-breeding worth the name between these two groups, hybrid females have a poor chance of being allowed to grow up. Pure Airedales have their value for certain definite purposes and a mating between two Airedales will produce dogs that fit certain requirements. The same is true of purebred Grey-hounds, the hybrids on the contrary have an uncertain value, they are not wanted.

We could give several instances where wild species coexist in one locality, which give perfectly fertile hybrids, in those cases where the two happen to fit into somewhat different ecological niches. The reason we choose the example of the dogs is to show, how in this matter of the remaining distinct of species, we are concerned with the relation between causes reducing the potential variability as opposed to frequency of crossing, whereas it is wholly immaterial what the nature of these causes happens to be. Geographical isolation keeps a group of organisms separate from other species, but geographical isolation is not a factor we are concerned with in the instance of our two dogs. In this case the breeders of the dogs, who control the choice of mates, sufficiently influence the proportion of inter-specific matings to intra-specific ones. This is a case where it is not only the constitution of the group, but clearly also the situation which makes it a species. It is not the nature of the different factors which reduce variability or which produce it, with which we are concerned, but wholly the proportion between these two groups of factors.

In certain species the factors which produce a preponderant number of intra-specific matings as compared to crosses, are to a great extent social factors.

This is notably true for human group, human species, but we saw it is equally true for some domestic animal species.

A clear understanding of the status of mankind would obviously be of great importance for a dispassionate discussion of questions of a political nature. There exists a regrettable lack of coöperation between Political philosophy and Genetics. Genetics, as far at it has interested itself in man, as Eugenics, has consistently concerned itself with the inheritance of details of qualities of individuals, with the inheritance of epilepsy, of supernumerary fingers, and has hardly ever discussed the influences which cause the *grouping* of individuals. It has looked upon such causes as upon things which do not affect the evolution of man, and has left a study of them to political science. Political philosophy, on the other hand, has been too prone to overlook the possibility, that within a nation there could be discoverable genetic differentiation not only, genetic differences between individuals, but differences between group of individuals.

And just as the foundations of Eugenics are failing, where they should include a study of the causes, environmental and other, which underly grouping of individuals, social science lacks a necessary foundation if it starts from the assumption that a people consist of fundamentally like individuals, without attempting to verify the point.

There ought to be no room for the two opposed views, that heredity, or that environment is mainly responsible for the qualities, the characters the development of a man or woman. The fact, that such a difference of opinion exists, should induce us to find a way so to state the problem that it be omes amenable to scientific, if possible to experimental investigation. The main point is not, whether an individual's qualities are more determined by heredity or by circumstances, but whether there is not some way to discover if the endless variation in man is simply caused by a combination of a great fluctuating variability in the genotype and a great variety of circumstances, opportunities, or whether it obscures an underlying, but discoverable grouping, caused by things which tend to specific diversity. In the present state of relative ignorance about these things, it is difficult not to be influenced by one's political opinion, clearly it should be our object to find a way which would lead past opinions and toward facts.

We are not greatly concerned with anybody's opinion, conviction, that all men are born equal, and should therefore have equal rights, or with the conviction of another man or group of men that they should rule a numerically greater group of men, because of a special innate fitness, hereditary superiority. We want to know, whether there are in man groups of individuals which are so situated and so constituted that the potential variability of those groups tends automatically to reduce itself, in other words, species.

And we want to learn about the causes which bring about the evolution of such groups. Or rather, we want first of all to find a way of discovering such facts.

The species question is of so great an importance for political philosophy as well as for Eugenics, that there must be some good reason for the fact, that it has not been more an object of nvestigation by either science. I think that the vast vagueness of the species concept, and the diversity of opinion among evolutionists is mainly to blame.

We will later take up the question, whether there is a specific difference between what are commonly called different races. Here the differences are so obvious and so constant, that Zoölogists do not hesitate. We will first take up the question whether there is any possibility of specific diversity within one nation.

We know that two species can coëxist in one environment, under circumstances which admit of frequent intercrossing, and that such species may yet conserve their identity, even if the hybrids produced are perfectly fertile. We know that such tests as the "sterility of hybrids" test for specific diversity are inadequate. If we refused to see distinct species where fertile hybrids are produced, there would be large groups of animals and plants, each comprising many true-breeding forms of great diversity, in which no species could be distinguishable, such as the great diverse group of the cattle, Bos, or the surface-feeding ducks, Anas.

Two species can coëxist in one environment, if only there is a multitude of individuals of each species, and if conditions are so, that the number of intra-specific matings remain far below that of the inter-specific matings. We know, that the tree-rat and house-rat in Java, and probably in North Africa inter-breed wherever they meet. The fact that there are localities where there are houses, habitable from a rats point of view but no good trees, and also localities where there are suitable fruit-trees, oilpalms, dates, coconut-palms, but no houses, keeps these two rats separate. In the villages, where good houses and good trees almost touch, hybrids are constantly being produced, but these are so far in the minority as compared with the house-rat millions and the tree-rat millions that they are swallowed up in the multitude. We can calculate how small are the chances of hybrids to affect the variability of a species into which they merge, and we have discussed the fact that a certain number of animals do not have an equally large number of parents, but a smaller number, and that in their turn, only a few of their number will be parents. Automatically a species purifies itself.

In plants, the case given by Bateson, that of Lychnis diurna and Vespertina, is a representative one. Hybrids are constantly being produced, and they always disappear, the old two species assimilate them.

Each species prefers a somewhat different habitat, but the difference, though efficient in keeping the majority of individuals of each species in localities where they do not meet individuals of the other species, does not preclude great numbers of both species to coëxist in the same environment, where it is neither too dry and sunny for vespertina, not too shady and damp for diurna.

In other cases, we find two or three or more species living in the same environment, when they are equally well adapted to this environment, and when the occasional hybrids produced are perfectly fertile. In all these instances we can find some influence which makes intra-specific matings very much more frequent than inter-specific ones, e. g. autogamy. And we must always remember, that the nature of the cause or causes, which bring about this preponderance of intra-specific matings, is immaterial. The species kept apart by it, are equally good species whether the cause is a geographic barrier, an adaptation to a somewhat different environment, a differential mating because of prejudice, or the interference of man. The different species of dogs in one city, the different species of cattle in one country furnish beautiful examples of groups of species which coëxist in one environment, which occasionally cross and yet preserve their identity as species. There is no fundamental difference between this cause of specific diversity and a geographic barrier, the so-called breeds of cattle are as good species as the tree-rat and the house-rat.

A variety is a number of individuals which all differ from the species to which they belong in an identical point, without having continuity. The Holstein breed of cattle constitutes a species. The occasional red and white individuals produced by it constitue a variety. Oenothera biennis is a species. The pale-yellow individuals it sometimes produces are a variety. Varieties can become species, if there is something in their make-up, or in the circumstances which make matings between its members more frequent than matings of its members with individuals of the species. Therefore, in self-fertilizing plants a variety is an incipient species. Self-fertilization saves it from going under into the species. But generally speaking, if we except self-fertilizing organisms, varieties do not become species. Species arise in a different way, in any way by which a group of organisms is split off from random mating with the whole group to which it belonged.

The most efficient cause for the reduction of variability in a species is not selection, but the fact that few individuals have a great number of descendants and many individuals have no descendants at all. This reduction of variability is very differently effective in different species. If we know that the number of tapeworms of a given species remains approximately constant from generation to generation, if we further look into the dramatic sequence of lucky, rare coïncidences which are together necessary for a single egg to develop into a mature worm, and if we count the millions of eggs produced by one individual,. we can see how amazingly rapid the process of specific purification must be in these animals, and how very ineffective occasional crosses must be here. If, on the other hand, we compare the number of off-spring produced by one African elephant to that produced by one tapeworm, if we compare the infant mortality of the two animals, we are impressed by the relatively small powers of automatic reduction of variability inside the species, and the relatively great effect of a hypothetical cross in the Elephant. In other words we see, that the barriers that keep apart species in such organisms as the elephants and the tigers must be more effective than the barriers, that keep species of flies and elm-trees and salmon apart.

How much inter-marriage is possible between two species without loss of identity? How effective must be the barriers between them? We see that this depends upon the power of the species automatically to restrict their variability. Hybrids will mate with pures, and their offspring with pures, until no trace of the cross is left. So long as the heightening of the variability by crossing does not exceed the possible reduction of variability through any or all of the causes which bring this about, a species will conserve its identity.

Whenever we compare specific distinction in man to specific distinction in animals and plants, we must remember that, in nature and under cultivation, both genetic differences and circumstances keep species distinct. We saw how in nature two species, mutually fertile, can live in approximately the same environment, if their difference in germinal constitution makes matings between members of one group more common than matings between individuals of different groups. The very existence of a multitude of individuals of one species guarantees the future of this species. And in those cases where a great difference is caused by a hereditary difference of genotype, the process is obvious.

Geographic barriers are the most striking examples of non-genetic causes for specific distinctness.

Under cultivation we again meet both causes. Here we find instances where groups of animals and plants constitute species in every essential sense, because the massing of a multitude of like individuals is a guarantee for the future of the type, even if no genotypic difference is responsible for the massing. Even obligatory allogamous plants like beets, maintain specific distinctiveness even where some cross-breeding between types takes place, when plants of one species are cultivated in masses. In man, we find all sorts of causes responsible for specific difference, specific distinctiveness.

Difference is a necessary result of distinctness. If two groups are effectively separated, each will tend to become pure for its own type, and chance will greatly determine what the type will be. It would be very remarkable if they finally developed in the same way, became pure for the same type. Two groups of people may originally be of common descent and keep from mixing because they live in different parts of the world. Each group will tend towards its own type and as each assimilates different smaller groups of original occupants or of immigrants there is enough chance for an eventual difference in type. (Canadians-Australians).

If groups of people living in close proximity, making up one nation, are sufficiently different to make the genotypic difference apparent, they tend to remain separate because of an aversion to mixed marriages which keeps these below the maximum. This maximum of the percentage of mixed marriages of the total number of marriages consistent with specific difference must in man, where the rate of reproduction and the infant mortality are so low, be necessarily very low. We see such an aversion to mixed marriages keeping species separate in cases where the genotype of the two species is very different, Negroes and whites, Chinese and Malay. Occasionally such differences are accentuated by differences of religion or language.

Next we have to consider the case in which originally non-genetic differences are strong enough to keep groups of people separate, even where they occur intermingled. In man, with the importance of language and of the printed record, the influence of outstanding individuals becomes important out of all proportion to their hereditary power, and quite independently from this. Apart from this influence of men upon other men, we have a circumstance which has great influence in man: the transmission from parents to off-spring of non-genetic developmental-factors. In the first place, we for the first time, meet here imitation and tradition as influences, which shape the development of man to an incomparatively greater extent than in any other animal. Imitation and tradition makes children very much more like their parents than an inheritance of genetic factors alone would make them, and more unlike members of a foreign species.

Then, also, the very environment is transmitted. In animals and plants and the majority of men every individual makes a new start, has to find a place for himself. In the case of man this means finding some suitable occupation, some means of getting his living in the world. Now in man it is possible for an individual who has found a place in the world, who for instance has come to use a certain extent of land on which to grow what he eats, not only to keep possession of that land for his own benefit, but to extend this privilege to his children. The other men will stand around and will not only let these children take possession, but they will help them defend their ownership. In other words they will help to keep other individuals from making this environment theirs.

This custom of allowing the privilege of use by an individual of a portion of the earth's surface or of materials, animals, to extend to his children is perfectly logical where man is as rare as some of the larger mammals. If the possession of some territory and tools and domestic animals by one family does not hinder anybody else from possessing his own garden and his own pigs, nobody objects to inheritance. But if it does, and unlimited accumulation of property and especially the privilege of holding accumulated property after death seems to produce a state of affairs where great numbers of individuals are without gardens and pigs, it is not to be wondered at, that some of these individuals begin to object. And we cannot be surprised to hear the statement, that unrestricted privilege of inheritance can not coëxist with democracy.

There is no reason for objection to a privilege of holding and transmitting property in land and material resources, in places where there is more than enough to go round. In newly settled countries like the United States half a century ago, and New South Wales to-day, there is no objection to land-holding and inheritance of land, because nobody is thereby debarred from going a little further into the woods and taking possession of another patch for himself and his children after him.

The question of rights in matters of privilege is intimately bound up with the species question as concerning human kind.

Nobody seriously considers the rights of the trees that grow on a certain spot, if he wants to cut them down in order to grow wheat for his family or clover for his pigs. And buffaloes and rabbits will be ousted from pasturages where they have been thriving for untold generations, if man needs the pasture for his live-stock. The interests of man, of our likes, come before the interests of all the other creatures.

But this feeling of solidarity with our likes, with people like ourselves is altogether more exclusive, than a feeling of the rights of men as against those of pine-trees or rabbits. In the first-place, settlers in a country already inhabited by men, may quarrel among themselves about rights in land and other property, but if the older inhabitants are visibly very different from themselves, the settlers will not seriously consider their ownership. Unless the aborigines have the organization and the temperament of the Zulus to resist infringements of their rights, these rights are by the settlers considered in approximately the same way as the rights of the pine-trees and the rabbits. For sentimental reasons reservations may be set aside for them, but this happens in the same spirit in which reservations are set aside for Buffaloes and redwoods and Egrets. The rights and privileges of our own "kind of people" come first everywhere. And that this spirit of solidarity with our own sort is not vague and general, is best illustrated by those instances where a few people of one nationality acquire rights in property, in diamond-mines or coal-fields situated in another country. Whenever such foreigners do not merge into the people among which they live, if they do not naturalize themselves, they sometimes feel hampered by having to submit to the laws of the country. The people in whose territory they live, may want to tax their property, or take away their privileges unless they assimilate themselves, unless they become citizens. In such cases the country-fellows of these emigrants will almost certainly sympathize with them. They will feel that the people

of "their own sort" have rights above other people, even if
they choose to live in the other people's country. Wars have
frequently been waged over such difficulties. Or rather, let
me say, very often it has been found possible to base the sup-
port and the sympathy of a people for a proposed war on this
feeling of solidarity irrespective of other peoples rights.

A commisioner of our Government had to travel in some
haste to a very small and remote and unimportant island to
investigate a report about trouble. He found the few hundred
inhabïtants considerably agitated, and two poor Chinamen
frightened out of their wits, barricaded in their little shop
which they had recently put up in the village. He found that
the villagers had attempted to kill the Chinese. By question-
ing, he found this state of affairs: Were they disapproving
of the shop? No, they thought the shop was wonderful and a
matter of considerable local pride. *A-ny-thing* that was made in
the whole wide world was assembled right here in this shop.
They had real round mirrors, and velvet skull-caps, and steel
fish-hooks, and papers of pins and fire-crackers and ginger-bread
and fine dried fish, no, they would not be without the shop
for anything. Had the Chinamen misbehaved? No, they had
not, they were very useful, they had already made them a
market for their cocoa-nuts and they had shown them how to
prepare bêche de mer in such a way, that is became worth
real money and produced striped silk handkerchiefs and
knives. Had the Chinamen insulted them? No, they had not,
but they were different. We people of the island own our trees,
and we have each our boat, and we fish in the sea and plant
rice in our clearings and shoot birds and build our houses in
the forest, wherever we please. These men are not our brothers
or our uncles, they can therefore not build a house here and
have their pigs run with our pigs. We thought that killing
them was the best we could do, for already our young men
were fighting and quarreling about them. In other words, the
islanders had their first taste of the immigration problem.

Mr. Colyn settled the difficulty in the following way. He

invented some impressive ceremony involving the drinking of blood, and he made the brothers be solemnly promoted to brothers and sons of the islanders, by paying two pigs and a box of tobacco for the privilege. Next he married them on the spot to daughters of a prominent citizen, and made them pay another pig for that privilege. And when the festivities were in full swing he sailed away contented and had never to come back there again.

A feeling of the specific difference between ourselves and strangers in our country is the cause of more hard feeling than economic objections to these strangers. And this is the more true, when we are ashamed of our feeling of superiority or of difference. What makes people ashamed of avowing, that they do not want great numbers of Chinamen or Hindus, to come into their country because they are too different to be assimilated? Partly it is due to a lack of insight, but mainly, I think, to a professed or real belief in the brotherhood of men, which makes people try to find the reasons for their aversion in a fear of the economic disturbance which the immigrants will cause.

We often read, that there would be no objection to the immigration of Japanese in California, if Japan at home had labour-unions and a resulting high standard of living and of wages. Some authors emphatically deny that "race" is at the bottom of the aversion, these people simply corner our labour-market, they will oust our people from their place in industry and trades.

All over the East-Indies we meet an objection to Japanese immigrants, which, curiously enough, is explained as to be due to the opposite reason, namely to a fear that these people will draw to them all the trade and capital. This is the objection made to Jews in Russia and to Arabs all over the East.

The real cause of the objection to such immigrants roots in the feeling of strangeness. These people do not mix well. They are not of our species and do not want to become of our species, they are a menace to a real unity of the nation, to the democracy which is so popular nowadays.

If we leave the question of immigration for the present, we are confronted with the following question: Are we, within the boundaries of a nation always concerned with one species in the biological sense or is it possible that more than one species of men coëxist in one nation?

From a political standpoint an answer to this question is of the very greatest importance, as we will presently see. This question is not to be decided by opinion, it is a matter of research. It would be possible so to arrange an enquiry, as to find out, in how far there exists within a nation classes of people which are separate species, classes within which marriages are so much commoner than marriages between members of different classes, that they assume the rank of species. And it should not be superlatively difficult to find out, what cause or combinations of causes are most effective to bring about this great preponderance of intra-specific marriages. In this connection we must remember the slow rate of reproduction and the small infant mortality in man, which makes it probable that only barriers of a certain magnitude, or combinations of barriers which assist each other's action, can bring about or continue a specific diversity in man.

How do specific differences within one nation originate? If we do not consider cases in which two different peoples live together within the political boundary of a country, cases in which such peoples tend to segregate, especially if they use a different language, the best examples of species within a nation are furnished by castes. Are castes within a nation of such a nature that their differentiation can be inferred? In several instances it is clear, that the specific difference between castes in a nation is not brought about by any cause or set of causes differentiating what once was one species, but by immigration. Immigrants of one species coming into a country in numbers big enough and with a strong enough tendency to marry with their kind will not merge into the older inhabitants. We have the case of the Manchus in China, the Hindus in Java, and, although the immigration was a thing of compulsion, the case of the black slaves in the United States. In some of these cases an incoming species may absorb certain elements from the older population, which formerly had not the status of a species. Such processes are not only matters of history and historical anthropology, but of everyday occurrence. (Japanese in California, Javanese in the West-Indies). Sometimes the former occupants will go to the wall. If a vast country is inhabited by a sparsely sown population of hunters, and an agricultural peope elect to discover and annex this country, these latter will be able to multiply up to the point, where they will drive the original hunters to agriculture and either to extinction or to the status of a lower caste.

Apart from immigration, castes may be formed by catastrophes, which force a group of people down to a level from which they can not rise again. In this way the Tan-kai people who live on ships around Shanghai and Hong Kong got differentiated. Slavery makes a species of the group of people subjected to it. On the other hand immigrants of a peculiar type, conquerors, may take possession of landed property which was formerly vested in the common people, and by the privilege of landownership hold their own as a species.

What are the barriers to specific unity within a people that can still keep species apart, or that can still become effective in differentiating a nation into species?

Religious opinion and tradition constitute one barrier. It seems probable that such a cause as the Reformation has done more than any other single cause, excepting alienation of common property by a privileged class, to cause specific distinctions within nations Religious tradition is a very common barrier to a merging of species that inhabit one country. It keeps Arabs and Chinese from merging into one species even in countries, where they are both immigrants and where they tend to the same occupations, landlord, tradesman, shoemaker, as in Java. Religious tradition keeps a scattered people without any natural centre of habitation into one species. Almost every country has its Jews, one species to every country.

More important as a barrier between species, and as a cause for the specific differentiation than religious tradition, and ranking almost with great physical differences which cause a mutual aversion, ranks the complex of causes which is peculiar to man and which we might call the privileged transmission of the material factors in individual development.

Not every man starts in life with the same opportunities for self-development. Some young people not only have the traditions and culture of their well-educated parents to start with, but these parents provide them with the means of a free development, give them the economic independence which frees them from giving at a low age all their best energy to a mere getting-on in the world. Such people are able to choose their life according to special ability much more effectively than those individuals whose parents cannot provide for them. This freedom to give relatively more time to reading and thinking and learning, to what we call culture, has a great influence upon the men and women who have the good fortune to receive it. It sets apart these people from the uncultured as effectively as the use of a different language does. We cannot be far wrong if we state, that to the cultured Englishman, the English miners are much more foreigners than the French people are. Culture and money, of which it is often a result, are effective barriers between groups that can be species or develop into species.

The inheritance of property, landed estate for instance, from parents to children, in the first place tends to split up a nation into at least two component species, a cultured, possessing species and a non-cultured and non-possessing one. It is obvious that this need not be always true. If there is enough to go round, we may have a state of affairs in which the accumulation of property by any of the people does not hinder others from doing the same. In a newly settled country the holding of land and the deeding of land to children does not hinder anybody from going a little further and after some trekking to pick another estate to develop for himself and his children. But it is very evident, that, as soon as through inheritance the growth of estates, which now can continue through generations, has reached a certain point, there is not enough for everybody, and some people will have to do without land or other property and the economic freedom for which such property stands. Through a continuance of the inheritance privilege estates tend to grow to alarming proportions, and consequently the owning class will tend to become a minority, a numerical minority.

But, through the fact that property makes possible self-development and learning and thinking and culture, such a class, which will be a species as truly as the meadow-lark or the Airedale-terrier is a species, wil tend to dominate politically. And all of these processes are automatic, and certainly not based upon conscious scheming, or on a wish to dispossess the other species, on the part of the cultured class.

Politically, it is of the utmost importance to know in every instance, whether a nation consists of one species or of several. Upon this knowledge largely, depends whether, from the viewpoint of each of the composing species, an existing system of government and law-making is fair, right. And it is clear that, unless in this respect the conditions of two countries closely approach each other, it cannot be thought of, to adopt in one, the political system of the other. In nations that consist of one species, speaking biologically (if such exist), the variability may yet be rather great. We know that variability within a species is common. Varieties may be produced frequently. In this way, we may have at any one time a number of aberrant inferior individuals, differing in moral or physical characteristics from the type of the species. These are the varieties with whom Eugenics has almost wholly concerned itself, neglecting for a study of the inheritance of abnormalities the causes of grouping of individuals, the study of evolution in makind. At the same time, we may see a number of superior causes of grouping of individuals, the study of evolution in makind. At the same time, we may see a number of superior individuals, men or women of genius, of special ability, who, from a genetic point of view may represent varieties just as the pug-nosed and the red-haired people.

In such a nation, almost any group of individuals is representative If the business of government and of law-making is detailed to a number of people, chosen for their mental capacity and ability for grasping problems, there is no danger whatever, that they will not further the interest of the whole people. They are, collectively, not different from the people, their interests are the common interests, their very mentality and view of ideals of life are those of the whole nation By leaving the government to a number of outstanding men and women, those individuals will not probably be each typical, and they will rather include all sorts of different varieties But they will counterbalance each other's deviations. The only form of government which would not be safe in such a nation would be a dictatorship, where the interests of one, possibly aberrant individual will not be identical with that of the whole people.

Everybody likes to have a voice in his own destiny and in that of his children. Everybody looks upon his children and upon his grandchildren as upon people fundamentally like oneself — identical: and everybody assumes that the same likeness as between parents and children and between great-grand-parents and great-grand-children exists between himself and the people around him. One wants to unite with his likes, together to regulate the affairs of common concern, to enact legislation, which will affect descendants as well as neighbours. Or if one cannot find time and leisure, or does not feel up to the requirements of actively participating in the bu iness of government, one wants to detail one's part in the ma ter, one's voice, one's vote, to somebody who is felt to be essentiolly like oneself, to an equal, biologically speaking, to a member of one's own species.

The idea of having people of a different species arrange one's life and that of one's descendants, is repulsive in the extreme. The recognition of this aversion is shown nowadays after a conquest of territory. The conquering nation hastens to do either of two things, bringing the inhabitants of the annexed country into the species, by compelling the younger generation to use the conqueror's language, and by favouring an emigration of country-people into the new lands, as Germany did in Alsace and Lorrain, or by giving them self-rule at once, as England did with the Transvaal and the Orange-Free-States republic.

If we have sufficient imagination, we may try to imgine ourselves in the place of a species of dogs in a world, where dogs arranged things to their taste and advantage. If we do so, we see that to an Irish terrier it must be extremely repulsive to have his affairs regulated by a group of collies, and reversely, that a collie would like to settle the matters of his state with a committee of other collies, and would be extremely reluctant to have some terriers, whose views on such very important subjects as rats and sheep are so very unlike his own, sit in judgment upon him and his race.

We do not believe, that it is very common for people to want to arrange matters for those, who are obviously not of their species. The missionary spirit, we hope and believe, is as much an aberration from normal as the genius for inventing a new dance. Those colonizing countries who are experienced, know better than to interfere very much with the customs and laws of the people they dominate, so long as these do not interfere with the happiness of the other groups. The English are content to let the Mahommedans in Ceylon be Mahommedans, and eat sheep instead of pigs, and let the Chinese eat both, and the Buddhists neither. They only interfere, when the Buddhists are going to knock the Mahommedans on the head for eating sheep, and start to loot their shops, as they were doing, when we were there. The governing colonizing people simply arrange things in as far as they materially affect the well-doing of their own class, and they will allow the Young

Men's Buddhist Association to have their grounds and buildings next door to the Young Men's Christian Association, so long as neither of them starts a row over the wall.

But a very much more serious situation exists in those nations, where the existence of two classes is not recognized.

Two classes A and B may coëxist within one nation, and marriages between members of the two different species may be frequent enough to give a false impression of unity and yet rare enough to keep the two separate as classes. In such a nation the upper classes, the class which monopolizes property to a great extent, and the resulting culture, is sure to dominate, and it is hardly fair to blame its members for arranging things in the way which looks most logical to them. But the other class, which is always a majority in densely populated countries, by the fact that inheritance of property makes its accumulation continue for more than one generation, does not really have a voice in matters of governing or law-making. And when a great many intellectual people begin to feel that they belong to this non-possessing class, which is bound to happen with an increase of the facilities for study and such inventions as that of the printing-press, this lower class, which constitutes a numerical majority, will find out that it has no hand in the arranging of its own affairs. As soon as this happens, there will be resentment and a wish to alter conditions. And as far as we can see, in such a case, two alternatives only seem to present themselves. In the first place, they may think of a system of government which will be representative, in such a way that the two classes which make up the nation will be represented in the same proportions in which they are present in the nation.

The other alternative, which will occur to them, is a radical change of government. They may want to overthrow the rule by class A altogether and start ruling the nation themselves, monopolizing the business of governing and law-making. This is what has been happening in Russia.

It will probably not occur to them, that there is a third course

This third possible way of so arranging matters of Government that they will be acceptable to all the people, consists of making one class of the nation. The barriers which keep apart two sections of mankind within one nation, are obviously not geographic barriers or sterility. Ranges of mountains will keep species of rabbits separate and species of butterflies, but man tunnels them and constructs railroads through the passes. Rivers will keep apart species of deer and of snakes and violets, but man bridges them.

The barriers with which we are concerned, are almost wholly social and therefore, amenable to change and removal. Two active causes which bring about the discrepancy between the number of inter-specific and intra-specific marriages in one nation, are the things which we call property and culture.

In certain circles and peoples, the recognition of these barriers to the unity of the nation, to democracy in other terms, engenders the wish to abolish property and culture.

However, if we examine the difficulty a little closer, we see that it is not property and culture as such, which act as barriers to unity, but the possibility of monopoly of property and culture by one class.

We see examples of communities where there is no inheritance, and we see how in these communities individual merits bring men to the top, and the needs of the needy are provided for and everybody has an equal chance of developing. We are not speaking of very desirable communities, because in desirable communities inheritance will soon play a rôle. Two instances may be given. We see such conditions in places where there is no property, or only vestiges of it, where very few aborigines live naked in a rich country that provides plenty of roots, and birds, and fat grubs, and fish, for everybody, where a man does not specially have to provide for his children, not because the state will set them up, but because nature does. The blacks of tropical Queensland are a case in point.

We also see communities without inheritance in those cases where the lfie of the community is of so ephemeral a nature, as not to include many cases of death and inheritance, even with an abundance of property being developed by the more industrious and able. A good instance is furnished by a rich gold mining ca p.

So long as a nation consists of two classes, there is bound to be some injustice. The two groups are too much intermingled, and their interests interact too much, to make a special government and a special set of laws possible. Unless the species differ in some marked characteristic, such as colour, a system of representative proportional government cannot succeed, because the exact proportions in which the species exist cannot be determined.

The negro-problem in the United States is a good example of the difficulties which arise, when two species coëxist, which are easily seen to be different. The abolishment of slavery in a certain sense created the difficulty. So long as the blacks were slaves, they were not citizens. When slavery was done away with, the slaves should have been done away with, we must see now that the best thing that could have been done to end the slavery, would have been to send the blacks where they came from. They are undesirable citizens in that they cannot mix. The whites do not want them to mix. They are a stumbling-block in the path of real democracy, for it goes very much against the grain to let them participate in government, not so much because the whites feel superior to the blacks, but because they feel they are different. The blacks in Tahiti have the same objection to letting the whites participate in government, not because they feel better than the whites, but because they feel that they are not of their species. And it is galling to an otherwise remarkably homogeneous people, to profess democracy and the brotherhood of men, and to stand for the choice of being consistent and admitting the blacks to a real equality, against every feeling of inequality, or of being inconsistent and of discriminating against them. There may not now be an efficient remedy for the situation, but the real nature of the negro-problem should be understood, and in this way the situation teaches a valuable lesson, namely that there is in the immigration problem something more than a matter of economics.

No matter how trade-unions develop in Japan, or in China, or in Arabia, no matter how high the standard of wages may become in those countries, the people of these nationalities are undesirable citizens for a country like the United States, because they do not mix, they remain separate. They cannot be taken up into the species. And real unity of species is a very great factor in the happiness of a people. According to the theory that economic reasons underly the aversion to immigration of Orientals, the Japanese aristocracy would furnish desirable immigrants. And the Sicilians who merely come over for a season, to work and earn a certain sum of money to take home would be undesirable immigrants. In reality neither one nor the other is true.

Are we, in immigration, concerned with the qualities of the men and women and children coming in, are we, for instance, to judge of their desirability according to one standard, rejecting individuals of inferior morality or intellect, and welcome sane, thrifty, healthy individuals, no matter of what nationality? Or should we remember that we are dealing with members of species coming in, and judge of the desirability of Chinese or Syrians, French, Poles and Fins as species, according to what we can find out about the way in which the Chinese and Syrians and French and Poles and Fins and their children and grand-children have assimilated themselves, and lost their identity in this common nationality, this common species? Bennett tells, how he felt the impulse of writing down "Yes" in the blank given him to fill in by the immigration officials of the United States, where it asked him "Are you an Anarchist?".

We felt the impulse to write across the blank "We are Hollanders," and leave the rest unanswered as immaterial.

How have we to consider the aberrations from normal type which we find occasionally at home, and among the immigrants? In so far as the aberrant individuals are found in one species and are alike in one characteristic difference, they constitute a variety. If among the Sicilians coming in one year, there are three red-haired ones, these, three constitute a red-haired variety. There need not be any family relationship, and it is probable that they will never have red-haired descendants. And we have to consider a colour-blind German, and an idiot Polish girl, and two Hungarian babies with hare-lips, as candidates for citizenship, who differ from normal Germans, and Poles, and Hungarians in varietal characters, in things which in all probability will have no continuity. The Eugenists have concerned themselves almost exclusively with the inheritance of such varietal characters, until it looks to the uninitiated that these things, which are certainly heritable, are important for the welfare of the nation, and as if any measure which excludes the genetically defective persons from procreating is necessarily beneficial and will do its bit toward an ultimate betterment of the "race". Underlying this idea, is a wholly erroneous conception of the real nature of specific stability, and of the effect of selection within relatively pure species. Species are pure when a very great majority is pure for a certain type. Small minorities have no chance, for the simple fact that their descendants have always again one normal parent, and will mate with normals and their children will mate with normals and so on. And we know, that there can be considerable crossing with other species without loss of specific identity. Colour-blind people have no chance to procreate their kind through generations, and bleeders may have an occasional bleeding grand-child, but the traces of their defect will have disappeared in a few generations. They will not affect the type of the species. On the other hand, the persons of musical genius, or of inventive genius, or of exceptional inherited ability of any kind, have no chance to heighten the status of the species into which they belong or merge. They have no future in this sense, just as little as the colour-blind and the feeble-minded have. A species is a remarkably stable thing, and for purposes of crossing one member of a species is as good as another. If somebody employs us to teach him how to make Holstein cattle out of his scrub cows, we can save him a lot of money by telling him to grade up his cows by repeated back-crossing, and for the first two generations or so, to buy bulls that have pure Holstein ancestry of the best quality, but who are to be had cheaply because of some varietal distinctive character, for instance to buy a red and white "cull" first and a blue and white next.

Mass immigration will certainly affect the species inhabiting the country where the immigrants come, provided they are "mixers". Especially if the immigration goes into a "new" country, where there are plenty of natural resources, and therefore good chances for getting on in the world, through personal effort. For in such countries some of the immigrants will work themselves up, and marry into the cultured classes, or marry their children into them, and some of the same origin will stay low down with the lowliest. Mass immigration certainly has tended to make one species out of the white inhabitants of the United States. And specific unity and democracy help each other along. The people of the United States have unity and they have a democracy. To outsiders, trained to observe in their own country, some measures which he sees go into effect, will look undemocratic, that is, he may feel that his people would not tolerate them. As an exemple of a small thing we might point to the obligatory wearing of gauze masks as measure against Spanish Influenza. But if one sees that the people do suffer arbitrary rules and do not chafe under them, one begins to marvel at the very great unity of the people, the very real democracy. For after all, real democracy is only real unity, and has nothing to do with freedom. It has been confounded with freedom only, through the circumstance, that it means freedom to hitherto unfree sections of a people. Rules of conduct will be popular and universally adopted if they correspond with usual conduct, if they are prescribed by essentillay typical individuals.

There is a happy conservative tendency in man to idealize

existing conditions and to believe that present conditions as we like them, are the result of our ideals coming true, rather, than that our ideals are the outcome of existing relations and conditions which have resulted somewhat independently of our efforts. This unity of the people of the United States, which is greater than any unity within a great nation of which we have knowledge, and which results in democracy, manifests itself in different ways, in great things and in small. In a negligible difference between the political platforms of the two great political parties, but also in the very striking absence of local types of architecture, in the fact, which strikes every European traveller, that all over the country, in Maine and California, the prosperous farm-living houses have the same windows, and the same gutters, the same roofs, even the same shade of sky-blue paint on the ceiling of the porch: and that the barns come in two colours only, whitewashed, and painted a hideous red.

This unity must be the effect of the enormous mass-immigration, which counteracted any dominance of any one people, which counteracts segregation into classes which can get specific rank, and makes a new, real species out of the mixed mass of humanity. These farm-houses are not English farm-houses, and those barns are not English barns by any means, or German barns, or Italian, they are American farm-houses and barns, and they are as truly national in type as the Belgian farms with the central barnyard, or as the enormous thatched buidings in Friesland.

On the other hand, we think that the democracy which is the natural result of the unity of a people, need not bring with it a real deep rooted wish for democratic ideals, brotherhood of man, and we believe that anybody who knows about the negro-question in the United states will have to concede the point.

When we study characters of men without keeping in mind grouping of men, species-formation in man, in other words, we place ourselves on the stand-point which the Eugenists have taken; it would seem, as if such an enormous immigration as that into the United States must make for diversity, and not for unity. And it is only when we have seen, that the essential nature of species is not purity, but eventual purity, automatic reduction of variability given in the constitution and situation, that we can understand why mass-immigration will make the nation which it affects into one great "Paarungsgenossenschaft," one great community, within which, random matings are the rule, and within which, there is only a slight tendency to specific differentiation.

We saw that the dislike against being ruled by a foreign species, is simply a manisfestation of the wish to regulate our affairs in common with those, whom we think are fundamentally as like to us as our parents and brothers and children are. Wherever men feel as in some European countries that a nation is composed essentially of two species, we see them become conscious of the possibility of injustice. Where a minority of privileged people practically monopolize the affairs of government, men are beginning to see that the present state of affairs is unfair to the other species, the numerical majority. They themselves in their turn, would dislike to see their affairs regulated by committees of workers and soldiers, and at the same time they are beginning to doubt whether the workers and soldiers will continue to submit to a government by the other species, and see important matters of state arranged by them. It becomes more and more obvious, that with an awakening of what is called "Class-conscious ness" but what we would prefer to call a "feeling of specific unity and specific disctinction", a reversal becomes probable. The majority, if beginning to feel that it has no voice in proportion to its numbers in regulating the law-making and governing, and feeling different from the ruling species, will want to overthrow things and govern in its turn. In this connection we must remember, that there is no real wish deeply rooted in any species of man, or in any great number of individual men (missionaries excepted), to meddle with the affairs of another species of men, in so far as these do not interfere with the interests of his own species. What most people want is to be free to regulate, in coöperation with individuals of their own type, the affairs which affect the community, and as soon as it is recognized that a nation does not consist of one species, there is a reluctance to interfere with the different species, and at the same time a reluctance against interference and regulation by the species which is not its own.

Revolutions will be effective in bringing the lower species into power, where there is such a distinction between two species. In countries where there is a real unity however, revolution is simply a symptom of discontent, but it will not result in anything like a difference.

When by a revolution the non-posessing species comes to power, it may do a number of very different things. It may set itself against culture, instead of arranging matters so, that culture will be attainable by anybody, and there is always some danger that such an unintelligent thing may happen. But where men of some education, and of clear understanding feel that they belong to this species, which is only another way of saying that they do belong to it, it is not likely that culture and capital are in much danger. In such a case it will be expedient to remove those things which hitherto have kept apart the two species, so as to make them into one. We have the example of the United States, where mass-imigration, mixing of the most heterogeneous white people makes one species out of the mixture. And the removal of the privilege of inheritance will in some countries be the removal of the main barrier. Such a result of a revolution will eventually tend to make for unity, and unity (which is not synonymous with equality by any means) promotes happiness, as it makes available all the best individuals for the business of government withou tany injustice to groups of individuals. It will make any group of individuals chosen for their individual merits, representative for the whole nation. There is, as far as we can judge, no real reason however, why this active removal of barriers to national unity should await a revolution, or an act of violence. It could clearly be done just as well by existing governments. If the ruling species in any country where there is no real unity sees, that the alternative to taking steps in the right direction is revolution, it may for instance set about to arrange for a gradual increase of inheritance tax, so that after a long series of years, in which a gradual adjustment is possible, inheritance is practically done away with. It may see that such a course of action would obviate injustice and hardship, and, colloquielly. "It would let them down easy." They may be sensible to do so, if only to obviate the very evident possibility of being thrown down abruptly. In the end, both ways of removal of this barrier to unity will come to the same thing, but intelligent regulation and slow change would obviate an enormous amount of suffering and economic waste. Even countries which now have unity, but whose unity of species is a result of a continued immigration, would be wise to study this point. For as soon as the country gets more settled, and natural resouces monopolized, and when at the same time the resident population increases, the immigration becomes relatively less important, some barriers to unity will eventually break up the nation

In every occidental nation nowadays, there is as trongly expressed wish for unity, a strong wish to be alike, and be able to coöperate and act together, and anything which conteracts this unity, is felt to be a hindrance.

This wish to be one expresses itself strongly in declarations of independence and nationality, in assertions of brotherhood of man, and in a tendency to resist all those causes which tend to split up a people The separation of church and state, more especially of church and school is a step in the direction of unity.

A strong, universal state-church may stand in the way of progress, and oppress science, but it certainly is a powerful factor in the unity of a people. As soon as there is any considerable dissention, however, prompt separation of church and state, church and school is best. Separation of church and state, eventually leads to the excessive splitting up of church denominations which we see in the United States. The more the churches become divided the less will be their power to keep apart classes of people, the better it will be for specific unity of the nation.

What is true of churches is true of parties in most of the essential points. Especially is this true where churches begin to go into politics. In this connection it is curious to note, that a church only begins to become a political party, when it gets into the minority. There is a very powerful Roman Catholic party in Holland, but there is no such thing in France.

In judging the effectiveness of party-lines as barriers to unity, barriers that keep species apart, we must not forget, that so far, only men have actively taken part in politics practically. So long as women are not deeply concerned in politics, party-lines are not especially effective as barriers.

There are, however, two sets of causes which will have to be examined here. In most of the European nations at least, parties, instead of following the lines set by the churches, are beginning more and more to confirm themselves to economic lines. And in addition to this, woman-suffrage has come at a very critical time, just when the curch has lost its power, just when growing syndicalism is emphasizing class-distinctions, and just when this last war has awakened whole nations to the fact, that they have no real unity, but that they each consist mainly of two classes (which we here have seen to be species in every essential), whose interests are not identical. Woman suffrage will certainly help to make party-lines more effective as barriers between species because it brings home to women the importance of party issues and it will make these party issues more than before a cause for selective mating.

To resume, there are nations which consist of only one species in the biological sense. In these, any group of men selected at random is representative. In these, popular vote is as good a way as any to decide particular measures. This unity of a people makes for real happiness. Other nations consist of more than one species. For so far as these are segregated geographically, they ought to separate and make two or more nations, that each have specific unity. In so far as they are not separated in this way, they can study the causes which keep apart the species, that make the number of inter-specific marriages remain far below that of the intra-specific ones. They should try to remove these barriers. Privilege of inheritance of property is one of the most efficient barriers. Its gradual removal would make possible an equal start, would provide a means for universal education and for effective state-help to disabled and old citizens. It would not abolish personal property and inequality, but it would make the possession of property a recognized sign of individual ability to do, to produce, to use.

Specific unity would tend to simplify politics, and do away to a great extent, with parties as they are now often differentiated.

Real, geographic isolation must act in man as in other animals, and produce local, circumscribed species. Several townships, island-populations are, or recently were, as effectively closed to admixture of foreign blood as groups of fishes in pools, or as the animals of islands.

Transportation breaks up such species. There may come a time when more new-comers get into a group than it can assimilate. A rather good test of the specific purity of certain groups of men, is given by the conservation of their local dress. In several townships in Holland, and several islands, the population rigorously conserves its dress. In some places so many new-comers have come in, and so many villagers have temporarily lived elsewhere and lost the habit of dressing in the way of their fathers, that the younger generation feels this dress as something uncomfortable, and gives it up. Those places where the railroad has come, almost certainly lose their local way of dressing. There are a few exceptions, in which a small town shows a decidedly agressive attitude towards outsiders, and has a feeling of superiority over its neighbours. (Huizen.)

Such people which give up a distinctive way of dressing,

show that they get from under the control of regulating factors, which kept the species together, just as stray and hybrid dogs get from under a similar control. What happens to them? They migrate to the cities more easily than the conforming members of the species. And the population of the cities has nothing to distinguish itself from the population of the other cities. The little islands of pure species all over a country melt away, they become smaller and smaller, and the sub-stratum in which they lay imbedded, is the more or less homogeneous residual population.

This mass of the population tends to dominate numerically, even if its birth-rate is not as high as that in the rural islands. Every time two or more local species coalesce, all those people come to belong to the residual species. For, as this group, this residual population of the country, shows no divisions anywhere, it is from the outside bounded only by people of different nationality, and by people speaking a different language. In such a group, the only division into species possible is the one on a basis of wealth and culture, or religion. If we speak of the French as compared to the Germans, we do not think of inhabitants of a Basque village, and not of Schwarzwalders, but of the French of Hâvre and Paris and Marseilles, and of the Germans of Hamburg, Berlin. People whose habitat is the whole of France, the whole of Germany. Common language and common laws hold these groups together.

Gradually, the local species in European countries are tending to merge into the great residual groups. The United States have been for all sorts of European communities, one vast international city in this respect. The only splitting up into species possible here, it seems to me, is a split such as has taken effect in the residual populations of the European countries. A secondary local differentiation seems little probable, in view of the transportation facilities.

BIBLIOGRAPHY.

ANONYMOUS author (Jenkins?). Origin of Species. North British Review, June 1867.

W. BATESON AND ASSISTANTS. Reports to the Evolution Committee of the Royal Society I, II, III, IV. 1902—1909.

W. BATESON. Mendel's Principles of Heredity 1909. Cambridge Univ. Press.

W. BATESON AND R. C. PUNNETT. The inheritance of the peculiar pigmentation of the Silky fowl. Journal of Genetics 1911.

W. BATESON. Problems of Genetics. Yale Univ. Press 1913.

E. BAUR. Vererbungs- und Bastardierungsversuche mit Antirrhinum majus. Zeitschr. fur Induktive Abstammung und Vererbungslehre 1910.

E. BAUR. Einführung in die Experimentelle Vererbungslehre. 2te Aufl. Borntraeger, Berlin, 1914.

J. BELLINGS. The mode of inheritance of semi-sterility of the off-spring of certain hybrid plants. Zeitschr. f. Induktive Abst. u. Vererbungslehre, 1912.

L. BONHOTE. Vigour and Heredity. Newman & Co. London, 1915.

E. BRAINERD. The Evolution of new forms in Viola through hybridization. American Naturalist, 1910.

A. CARRELL. Further studies on the transplantation of vessels and organs. Amer. Phil. Soc. Proceedings Vol. 47, 1908.

W. E. CASTLE. Some biological principles of Animal Breeding. Am. Breeders Magazine, 1912.

—— Piebald Rats and the theory of the genes. Proc Nat. Acad. of Science, 1919.

—— New light on blending and Mendelian Inheritance. Am. Naturalist. 1916.

C. CORRENS. G. Mendel's Regel über das Verhalten der Nachkommenschaft der Rassenbastarde. Berichte Deutsche Bot. Ges., 1900.

H. VON CRAMPE. Zuchtversuche mit zahmen Wanderratten. Landwirtschaftliche Jahrbücher, 1883—1884.

F. LE DANTEC. La lutte universelle. Flammarion, Paris, 1913.

CH. DARWIN. Variation of animals and plants under domestication. London, John Murray, 1905.

290 BIBLIOGRAPHY.

Ch. Darwin. Origin of Species. London, John Murray, 1899.

C. B. Davenport. Statistical methods with special reference to biol. var. 2d edition, New York, 1904.

—— The origin of domestic fowl. Journal of Heredity, 1914.

B. M. Davis. Genetical studies in Oenothera. Amer. Natur. 1910—1913.

J. Detlefsen. The fertility of Hybrids in a Mammalian Species Cross. Amer. Breeders Magazine. Vol. 3, 1912.

D. Dewar and F. Finn. The making of species. John Lane, 1909.

G. Dorfmeister. Uber den Einflusz der Temperatur bei der Erzeugung der Schmetterlingsvarietäten. Mitt. der Naturw. Ver. f. Steiermark, 1879.

H. Drinkwater. An account of a Brachydactylous Family. Proc. Roy. Soc. Edinburgh, 1908.

—— Inheritance of artistic and musical ability. Journ. of Genetics, 1916.

E B Droogleever Fortuyn. De cytoarchitectonie der groote- hersenschors van eenige knaagdieren. Proefschrift. Amsterdam, 1909.

R. R. Gates. Breeding-experiments which show that hybridization and mutation are independent phenomena. Zeitschr. f. Ind. Abst u. Vererbungslehre, 1914.

E. Gayot. Lapins, Lièvres et Léporides Librairie Agricole, Paris, (non datée).

R. Goldschmidt. Einführung in die Vererbungswissenschaft. Leipzig. Wilhelm Engelmann, 1911.

J. Grinnell. An account of the Mammals and Birds of the lower Colorado valley. University of California publ. Zool., 1914.

J. Grinnell and H. B. Swarth. An account of the Birds and Mammals of the San Jacinto area of South Calif.
    With remarks on the behaviour of geograhpic races on the margins of their habitata. Univ. of Cal. Publ. Zool., 1913.

J. Grinnell. Field-tests of theories concerning distribution control. American Naturalist, 1917.

E. Häckel. Vererbungslehre. 2e Auflage. Braunschweig, 1913.

S. W. Hardwood. New Creations in Plant life. An author. account of the life and works of Burbank. New York, Mc.Millan, 1907.

A. L. Hagedoorn. Mendelian Inheritance of Sex. Archiv. f. Entwicklungsmechanik der Organismen, 1909.

—— Autokatalytica, substances the determinants for the inheritable characters. Vorträge und Aufsätze uber Entwicklungsmechanik Roux. Leipzig, 1911.

—— The genetic factors in the development of the house-mouse which influence coat-colour. Zeitschr. f. ind. Abstamm. u. Vererbungslehre, 1911.

A. L. HAGEDOORN. Les facteurs génétiques dans le dévelopement des organismes. Bulletin Scient. de la France et de la Belgique, 1912.

A. L. HAGEDOORN and A. C. HAGEDOORN. Geneeskundig onderzoek voor het huwelyk. Tydschrift voor Geneeskunde, 1914.

―― Can selection improve the quality of a pure strain of plants? Journal Board of Agriculture, 1914.

―― Studies on variation and selection. Zeitschr. f. Ind. Abst. u. Vererbungslehre, 1914.

A. L. HAGEDOORN. Opmerkingen over fokken. Veeartsenijkundige bladen voor N.-Indië, 1915.

A. L. HAGEDOORN and A. C. HAGEDOORN. Parthenogenese by hoogere planten. Teysmannia, Buitenzorg, 1916.

―― Rats and Evolution. Amer. Naturalist, 1917.

―― Inherited predisposition for a Bacterial disease. American Naturalist, 1920.

J. HONIG. Kruisingsproeven met Canna Indica. Verslag Wis en Natuurk. Afd. Koninklijke Acad. v. Wetensch. Deel XXII, 1914.

H. S. JENNINGS. Heredity, variation and evolution in Protozoa II. Proc. Am. Phil. Soc., 1908.

―― Heredity, variation and the results of selection in the uni-parental reproduction of Difflugia corona. Genetics, 1916.

W. JOHANNSEN. Om nogle Mutationer in rene Linier. Biol. Arbedjer Tilegn. E. Warning, 1911.

―― Elemente der Exacten Erblichkeitslehre. Kopenhagen, 1913.

―― Arvelighed i. Historisk o. Experimentel Belysning. Gydendalske Boghandel, Kopenhagen, 1918.

A. JORDAN. Remarques sur le fait de l'existance en société à l'état sauvage des especes végétales affinés et sur d'autres faits relatifs à la question de l'espèce. Bull. Ac. franc. Avanc. des Sciences Lyon, 1873.

D. S. JORDAN. Factors in Organic Evolution. Univ. press Univ. of Cal., 1894.

A. LANG. Ueber die Mendelschen Gesetze, Art- und Varietätenbildung, Mutation u. Variation, insbesondere bei unseren Hain- und Gartenschnecken. Verhandl. der Schweiz. Naturf. Gesellsch. Luzern, 1906.

―― Die Erblichkeitsverhältnisse der Ohrenlänge der Kaninchen nach Castle und das Problem der Intermediären Vererbung und Bildung konstanter Bastardrassen. Zeitschr. f. Ind. Abs. u. Vererbungslehre, 1910.

―― Fortgesetzte Vererbungsstudien Zeitschr. f. Ind. Abs. u. V., 1911.

J. B. A. LAMARCK. La philosophie Zoologique. Paris, 1809.

B. E. LLOYD. The growth of groups in the Animal Kingdom. Longmans, Green, and Co. London, New York, Bombay, 1912.

J. LOEB. Chemische Entwicklungserregung. Julius Springer. Berlin, 1909.

J. P. LOTSY. La théorie du croisement. Arch. Néerl. Sciences exactes et naturelles, 1914.

—— Het tegenwoordig standpunt der Evolutieleer. Nyhoff, den Haag, 1915.

J. G. MENDEL. Versuche über Pflanzenhybriden. Verhandl. Naturforsch. Veriein Brünn, 1865.

A. R. MIDDLETON. Heritable variations and the results of selection in the fission rate of Stylonychia pustulata. Journ. Exp. Zool., 1915.

OTTO L. MOHR. Charater changes caused by imitation of an entire region of a chrômosome in Drosophila. Genetics, 1919.

T. H. MORGAN. The physical basis of heredity. The Lippincott Co.,1919.

C. NAUDIN. De l'hybridité comme cause de variabilité. Annal. des Sciences, Ser. 5, 1865.

F. H. NEWMAN F. Z. S. On some turtle-dove hybrids and their fertility. Avicultural magazine, 1904.

NEW REPUBLIC Editorial March 22, 1919.

N. HERBERT NILLSON. Die Variabilität der Oenothera Lam. und das problem der Mutation. Zeitschr. Abst. u. Vererb , 1912.

H. NILLSON EHLE. Om lifstyper och individ. variat. Botaniska Notiser, 1907.

—— Einige Ergebnisse von Kreuzungen bei Hafer und Weizen. Botan. Not. 1908.

—— Spontanes Wegfallen eines Hemmungsfactors beim Hafer. Zeitschr. Ind. Abst u. Vererb., 1911.

L. PASTEUR. Les microbes organisés. Mémoires de Tyndall et Pasteur, 1878.

R. PEARL. The mode of inheritance of fecundity in the domestic fowl. Journ. Expt. Zool., 1912.

—— On the results of inbreeding in a Mendelian Population. Amer. Natur., 1914.

J. A. N. PÉRIER. Croisements ethniques. Paris. (Non daté.)

J. C. PHILLIPS. Size-inheritance in ducks. Journ. Exp. Zool., 1912.

L. PLATE. Die Erbformeln der Farbenrassen von Mus musculus. Zool. Anzeiger, 1910.

P. S. M. PODMORE. The hybridization of Columba Palumbus. The Zoologist, 1903.

H. PRZIBRAM. Ubertragungen erworbener Eigenschaften bei Säugetieren und Versuche mit Hitzeratten. Verhandl. der Ges. Deutscher Naturf. u. Arzte Salzburg, 1910.

R. C. PUNNETT. The inheritance of coat-colour in rabbits. Journal of Genetics, 1912.

R. C. PUNNETT and P. G. BAILEY. On inheritance of weight in poultry. Journal of Genetics, 1914.

—— Studies in rabbits. The inheritance of weight. Journ. of Genetics, 1918.

A. J. QUETELET. Recherches Statistiques. Bruxelles, 1844.

W. RITTER. Marine Biology. American Naturalist, 1909.

F. M. ROOT. Inheritance in the asexual reproduction of Centropyxis aculeata Genetics, 1918.

W. ROUX. Die Entwicklungsmechanik. Leipzig, Wilhelm Engelmann, 1905.

R. N. SALAMAN. The inheritance of colour and other characters in the potato. Journal of Genetics, 1910.

G. H. SHULL. The presence and absence hypothesis. Amer. Nat., 1909.

—— Defective inheritance-ratios in Bursahybrids. Verhandl. des Naturforsch. Vereins in Brünn. Band 39.

B. J. STOCKING. Variation and inheritance of abnormalities occurring after conjugation in Paramecium caudatum. Journ. Exp Zool., 1915.

A. B. STOUT. The establishment of varieties in Coleus by the selection of somatic variations. Carnegie Inst. Publ. No. 218, 1915.

F. SUMNER. Continuous and discontinuous Variations and their inher. in Peromyscus. American Naturalist, 1918.

—— Geographic Variation and Mendelian inheritance. Journ. Ex. Zoology, 1920.

J. B. J. SOLLAS. Inheritance of colour and of supernumerary mammae in guinea-pigs with a note on the occurrence of a dwarf form. Report to the evolution Comm. Royal Soc. 1909 V.

E. VON TSCHERMACK. Notiz über den Begriff der Kryptomerie. Zeitschr. Abst u. Vererbungslehre, 1913.

R. HAIGH THOMAS MRS. Colour and Pattern-transference in Pheasant crosses .Journal of Genetics, 1916.

L. T. TROLAND. Biological Enigmas and Enzyme Action. Am. Naturalist, 1917. Vol. 51.

A. H. TROW. A criticism of the hypothesis of linkage and crossing-over Journal of Genetics, 1916.

PH. DE VILMORIN. Argémones hybrides. Revue Horticole, 1912.

L. DE VILMORIN. Notices sur l'amélioration des plantes par le semis. Nouvelle édition, Paris 1886.

H. DE VRIES. Intracelluläre Pangenesis. Jena 1889.

H. DE VRIES. Die Mutationstheorie. Leipzig 1901 und 1903.

—— Species and Varieties, their origin by Mutation. Open court publ. C. Chicago, 1905.

M. WAGNER. Uber die Entstehung der Arten durch räumliche Sonderung. B. Schwabe, Basel 1889.

A. WEISMANN. Vorträge über Dessendenztheorie. 3e Aufl. Jena, 1913.

H. G. WELLS. Mankind in the Making. Scribner, New York, 1904.

H. WINKLER. Untersuchungen uber Propfbastarde. Jena, 1912.

B. I. WHEELER. Letter on Japanese immigration to the Californian newspapers. September 1920.

T. B. WOOD. Note on the inheritance of horns and face-colour in sheep. Journal of Agric. Science 1906.